T0073650

Bound by Muscle

Bound by Muscle

*Biological Science, Humanism, and
the Lives of A. V. Hill and Otto Meyerhof*

Andrew Brown

OXFORD
UNIVERSITY PRESS

OXFORD
UNIVERSITY PRESS

Oxford University Press is a department of the University of Oxford. It furthers the University's objective of excellence in research, scholarship, and education by publishing worldwide. Oxford is a registered trade mark of Oxford University Press in the UK and certain other countries.

Published in the United States of America by Oxford University Press 198 Madison Avenue, New York, NY 10016, United States of America.

Library of Congress Control Number: 2022943843

ISBN 978-0-19-758263-3

DOI: 10.1093/oso/9780197582633.001.0001

1 3 5 7 9 8 6 4 2

Printed by Integrated Books International, United States of America

For Naomi, Poppy, Sabrina, Eddie, Elliott, and Des

Contents

Acknowledgments

The format of this book changed several times during the research and writing. At the outset, I was trying to link four near contemporaries from Trinity College Cambridge (William L. Bragg, Charles G. Darwin, Ralph H. Fowler, and A.V. Hill) into a book that would recount their experiences before and during the 1914–1918 war. This led me to spend weeks engrossed in the Churchill Archives Centre (CAC) in Cambridge as well as time at the Royal Institution, London. Allen Packwood, Andrew Riley, and the staff at CAC were helpful as always, and Jane Harrison at the RI supplied interesting material on Bragg and Hill. Suzanne Foster gave me on a tour of Winchester College and unearthed material about Fowler. His granddaughter, Mary Fowler, then Master of Darwin College Cambridge, provided hospitality and family information.

The plot for that book never gelled. One reason was that Hill led me into reading about the pre–World War I physiology department at Cambridge—less-explored territory than that occupied by the physicists in the Cavendish Laboratory, while offering figures of equivalent scientific stature. Angela Creager invited me to attend weekly seminars in the History of Science Department at Princeton, where I listened to some spellbinding presentations by graduate and doctoral students, often on biologists I had not encountered before. Hill's and Meyerhof's names came up indirectly from time to time, eventually prompting me to investigate Meyerhof, whose name was a dim memory from biochemistry lectures. I was unaware of the bonds between the two men, formed first through their researches on muscle physiology, strengthened by their joint Nobel Prizes, and made unbreakable by their resistance to the forces of fascism.

Unlike Hill, Meyerhof has a very small archive as a result of losing so many personal belongings during his wartime escape from Europe. I was surprised to find one box of wartime letters to his children in the University of Pennsylvania Archive, where Jim Duffin and his staff made me welcome. Paul Weindling and David Zimmerman, two distinguished scholars of World War II events, supplied me with valuable papers. Jane Rosen at the Imperial War Museum (IWM) allowed me immediate access to Sir Henry Tizard's papers during a visit to London. I am also grateful to staff at the National Archives, Kew. At the Royal Society, Rupert Baker provided me with a stack of records during lockdown; Jonathan Bushell and Katherine Marshall found photographs for me. Collecting photographs was often an adventure, never more so than tracking down Figure 13.1 portraying scientists at Woods Hole, eventually located by Heather Smedberg in a collection at UC San Diego. I was also aided by staff at the Wellcome Collection (Holly Peel), the IWM (Dave McCall), and the other collections listed in the individual photograph credits. The staff at the Free Library of Philadelphia obtained some obscure books for me, as did Jill LeMin Lee at the Athenaeum of Philadelphia.

Physiology in the first half of the twentieth century was a broader subject than exists now. Nancy Curtin, an emerita professor from Imperial College London whose career centered on muscle physiology, kindly read the manuscript for me. She has published an excellent, more technical, paper about Hill's contributions (Barclay and Curtin, 2022). I also benefited from comments by three professorial friends who are not muscle physiologists: H. Franklin Bunn, Jonathan Cole, and Lynn Penn. Graham Farmelo, Julian Loose, Latha Menon, Robbie Whitehouse, and Alexander Wolff provided useful nudges. I should state the obvious at this point: remaining errors belong solely to me.

Embarking on a new book reminds one about the loyalty and kindness of old friends, and it is a pleasure to acknowledge the Charltons, the Hineses, the Reads, some Shieldses, the Tacketts, Simon Lawday, Peter Phillips, and Michael St Clair. Above all, I would like to thank the wonderful Sharon Wallace for her forbearance during lockdown, when most of the writing was completed.

At the Oxford University Press in New York, my editor Jeremy Lewis was always responsive and positive. His former assistant, Bronwyn Geyer, could not have been more reassuring as she moved to a new position; Michelle Kelly, the project editor, is a model of clarity. My thanks also to Leslie Johnson, the project manager, and to Timothy DeWerff for his meticulous copy editing. Vinothini Thiruvannamalai supervised the book production by Newgen in Chennai.

Since embarking on the dual biography, I have enjoyed enthusiastic support from Hill's and Meyerhof's grandchildren. The book is dedicated to mine.

1

Breakfast at the Hills'

On Friday, 26 October 1923, Professor A.V. Hill started his day with an early morning run on nearby Hampstead Heath, followed by a brisk, cold bath. In his late thirties, Hill was the newly appointed Jodrell Professor of Physiology at University College London (UCL). He and his family had only just moved down from Manchester into their substantial home, Hurstbourne, on Bishopwood Road in Highgate. His young children were still adjusting to their new surroundings and were at breakfast with their parents. It was cooked and served by the cook-gardener who had been with them in Manchester; a nursemaid had already helped the children to dress. The three oldest—Polly, David, and Janet—were already attending a nearby private school, Byron House. Maurice, aged four years, would join them the following year. They were still excited about their relocation and discussed everyday observations and disappointments with their genial father.

Their mother, Margaret, sat at the opposite end of the table to her husband, reading that morning's edition of *The Times*. Her maiden name was Keynes. She grew up in Cambridge, where her father was the senior administrator of the university and her mother, Florence, was the town's leading figure for charitable works. Margaret's older brother, John Maynard Keynes, was the most famous and controversial young economist in the country, having just published *The Economic Consequences of the Peace* and resigning from the Treasury while attending the peace conference at Versailles. Her younger brother, Geoffrey, was a surgeon at St. Bartholomew's Hospital in London, where he was pioneering conservation surgery for women with breast cancer. Geoffrey's wife was one of Margaret's closest childhood friends, Margaret Darwin. Darwin's sister, Gwen Raverat, drew a fine portrait of Margaret Keynes as a young woman, showing a mass of curls surrounding an alert face—her lower lip was generous like Maynard's. A childhood friend described her as having the look of a bad angel.

Florence Keynes set up the Boys' Employment Registry (BER) in Cambridge to take boys away from casual work on the streets, where they were prey to exploitation and moral corruption, and find them apprenticeships. Margaret shared her mother's sense of social justice and noticed the burdens placed upon less fortunate families. So, it was natural for her to follow her mother's example as a charity organizer. She started by becoming the first assistant to another young woman at the employment registry, Eglantyne Jebb, whom her mother had recruited to a few years before. Like Margaret, Jebb was well connected in the university social scene. She was thirty-four years old and a pre-Raphaelite beauty—tall, with flaming copper hair, light-blue

Bound by Muscle. Andrew Brown, Oxford University Press. © Oxford University Press 2023.
DOI: 10.1093/oso/9780197582633.003.0001

eyes, and porcelain-white complexion.[1] Jebb experienced indifferent health due to a chronic thyroid condition; she left the BER in April 1908 so that Miss Keynes became the Hon. Secretary. Despite Margaret's forebodings, she grew into the position quickly and persuaded her family to make donations to support individual boys who won her sympathy. She and Maynard admired each other: when she wrote a report, *The Problem of Boy Labour*, he praised it as "extraordinarily good—so written as to be a most interesting and even moving document."[2]

Another reason for the closeness between Maynard and Margaret was that both had knowledge of the other's same-sex love affairs. Maynard had a prolonged relationship with the artist Duncan Grant, who once started a portrait of Margaret. She had quickly developed a crush on Jebb for her cleverness, beauty, and goodness; the attraction only increased after Eglantyne left the BER. The two young women still saw each other in Cambridge and spent a week together walking in Yorkshire in 1908. Jebb's mother became ill and she took her unmarried daughter with her as a companion to Switzerland. The two young women communicated by writing hundreds of letters that were full of emotion and sometimes passionate desperation. Eglantyne was depressed and isolated in Switzerland, and Margaret tried to reassure her: "You are my wonderful treasure that no one but yourself can take away from me."[3] The emotional states underlying such exchanges between young women in Victorian and Edwardian times were obviously complex and did not necessarily imply a sexual relationship.[4] An important difference from "the love that dare not speak its name" was that female homosexuality was never against the law in England. But by the end of 1911, whenever Eglantyne came back to England to visit, she and Margaret arranged to share a bed.[5] Their letters became more affected. Eglantyne became "My own darling Lulsy," while Margaret was "Polly," or "P" or "My Best." In April 1912, Margaret wrote to Eglantyne about her "frantic desire" to be together always, and the pair often mused about marriage (Figure 1.1).[6]

This was around the time that Margaret first met A. V. Hill, who was then a Cambridge physiology research student. He came to tea at the Keynes' home at 6 Harvey Road. Apart from his mother and sister, he had no real experience of the opposite sex and certainly no idea about the hormonally charged maelstrom he was skirting. In his innocence, he invited Margaret to take a ride in the wicker sidecar of his newly acquired motorbike, Buster, and then for a trip on the River Ouse. Margaret, who was used to invitations to lunches, parties, and balls, probably thought little of the suggestion and excused herself from going. Over the summer, he helped Margaret to craft a letter to be distributed to all new Cambridge undergraduates, warning them of the negative consequences of "indiscriminate charity": "the misplaced generosity of benevolent people" that led to "a spirit of dependence and a disinclination for regular work."[7] Mrs. Keynes invited him to a family lunch again in August and co-opted him onto her charity committee. The following month, Margaret was away on holiday with Eglantyne and Florence wrote to her to say: "*A. V. Hill came into tea today. He has just come back from Ireland where the yachting was not a great success on account of storms. … He is bronzed by the sea breezes and looks more like a naval officer than*

Figure 1.1 Margaret Neville Keynes (1885–1970) (left) and Eglantyne Jebb
(1876–1928) (right)

Photograph courtesy of Professor Nick Humphrey/Churchill Archive Centre, Cambridge (Papers of
A.V. Hill, AVHL).

ever."[8] Glorying in the escape to her lover, it seems unlikely that Margaret took her
mother's broad hint, even though it was juxtaposed to a list of recent engagements in
Cambridge.

AV, as he was generally known, did little to advance his cause with Margaret during
the Michaelmas term. Margaret and Eglantyne exchanged Christmas letters in which
they talked again about "marriage" and buying a house together in Kennington,
south London.[9] Eglantyne's mother was supportive, whereas Mrs. Keynes redoubled
her efforts to promote AV: she invited him for Christmas. Margaret mentioned to
Eglantyne that "Mother has a particular affection for A.V. Hill" without disclosing any
attraction on her own part. Indeed, she assured her Lulsy, "P wants to kiss you *vezzy*
badly." She wrote again on 9 January 1913, telling her darling that AV was going to
write an article on boys' employment for her and was coming to lunch on Saturday.
She thinks Jebb would agree with many of his views since he is "interested in the
Army without being jingo, Church but not bigoted, and a philanthropist—I saw he
contributed to the Balkan fund."[10] Three days later she informed her Lulsy: "A.V. Hill
came to lunch yesterday, we are all rather in love with him." And on 14 January, "*My
own darling Lulsy ... he is the most friendly person to all the world excepting faddists.
I like to have him as a friend ... _Much_ love from your own P.*"

Margaret mentioned to Hill that "her great friend, Miss Jebb"[11] would be visiting
Cambridge and she wanted him to meet her. Margaret was fast approaching her

twenty-eighth birthday, well over the usual age for starting a family. She knew enough about Cambridge to realize that Hill, as a fellow at Trinity College and recently appointed as a junior dean, was on the threshold of a promising career. He could be an ideal husband if she so chose. She did ask him how "anyone can devote themselves entirely to science who is not the lucky possessor of unearned income."[12]

Early February saw hectic developments, with Hill emboldened and Margaret suddenly prepared to consider matrimony with a male. He wrote on the first of the month suggesting to Margaret: "*we might promote a company for the advancement of humanity ... in which you and I will be partners, you to act as reformer of morals and I as tinker of apparatus.*" She was understandably baffled by this elliptical proposal and told Eglantyne, "*Of course I shall go and see what he means.*"[13] Poor Eglantyne must have known instantly—she had been becoming increasingly depressed by Margaret's growing affection for Hill. Margaret for her part, whether she realized it or not, was seeking Eglantyne's approval for her impending choice.

AV was never called by the first name, Archibald, and Margaret addressed various uncertainties in her reply to him on 3 February.

> I don't know what to call you ... It worries me to think how much too good you are for me because you may find it out too. I admire you very much—I don't know if I am in love with you but I should like to help you, to make you happy, to look after you. I should be unstable if I wasn't going to see you almost at once. Mother says she will be in tomorrow from 3 to 4. ...
>
> Forgive me if I have disappointed you this afternoon. I cannot tell you how much I cherish you. It seems almost impossible that you can really want me, but you are not pledged. We will "walk out" together and you can get to know my foibles.

It does not sound like an immediate, unconditional "Yes." The following day, he was making a more formal proposal in the drawing room at 6 Harvey Road, when Maynard Keynes inadvertently walked in. He captured the scene for Duncan Grant:

> Poor young man. He was practically in tears and had extraordinarily the appearance of having had his face bashed in. Margaret was really quite calm and collected. ... Tea had just been laid for two and their chairs were drawn by the fire. ... He was a pitiable sight and yet he's quite nice and really very suitable. Indeed it's not his fault that's responsible, but the fact that Margaret is more deeply entangled than ever in a sapphistic affair.[14]

Margaret broke the news to Jebb that AV had asked her father for her hand.

> My darling Lu,
> Of course father is satisfied after the interview. So he is going to tell his mother at once & I shall probably see her tomorrow.[15]

Maynard confirmed to Duncan Grant that the Keynes family was well satisfied with Margaret's choice, telling him that Hill is "frightfully puritanical—but then that suits her."[16] He added snidely, "even the sapphist not notably obstreperous." Margaret had told Maynard that Jebb was "much too sensible to make a fuss."[17] Indeed, Eglantyne behaved with great dignity over her rejection and resolved to make her own life. Margaret spent what must have been a difficult weekend with her and her mother in London. Eglantyne announced that she was planning a trip to the Balkans to assess relief work being carried out in Macedonia, which had just been divided between Serbia and Greece.[18]

Jebb returned in time to attend the wedding, which took place in Cambridge on 18 June 1913. The newlyweds rode on Buster to Scotland for their honeymoon on the Isle of Skye and returned in mid-August to a rented house overlooking Parker's Piece, which Florence Keynes had redecorated while they were away. Polly was born almost exactly one year after the wedding; two more children followed during World War I and Maurice arrived in 1919. Marriage to Hill transformed Margaret's life—she emerged from the war years, followed by difficult times in Manchester, as a much more self-confident woman.

As she was half-listening to the conversations at the other end of the breakfast table, Margaret noticed a report in *The Times* from Stockholm, dated the previous day. "Vivian," she asked, "Who do you think won the Nobel Prize for medicine?" Hill responded with the names of one or two senior physiologists he admired, probably including Sir Charles Sherrington, and Margaret said, "No, Vivian, you did!" The announcement stated that Professor Archibald V. Hill was to share the prize equally with Otto Meyerhof, a professor of physiology at Kiel University. Hill's half was for "his discovery relating to the production of heat in the muscle," while Meyerhof's was for demonstrating "the fixed relationship between the consumption of oxygen and the metabolism of lactic acid in the muscle."[19] Hill was of course delighted to win, and to share the prize with Meyerhof—a physiologist of his own generation, and one German whom he liked and respected. In the brief time before the children set off for school, their stunned parents tried to explain the significance of winning a Nobel Prize. On arrival at Byron House, Janet announced excitedly to the class that her daddy had won the Derby.

One week later, AV received a letter in the physiology department at UCL from Otto Meyerhof. It was typewritten in rather formal German and expressed his heartfelt happiness, both in winning the prize and sharing it with Hill.[20] Like most laureates down the years, he was also surprised. He was nervous that there was so much to do before they met again in Stockholm. More telling were the marginal notes and postscripts that he added in pen—remarks that would not be read by his secretary or other colleagues. He was amused that while most seemed genuinely joyous at his unexpected honor, he also detected an undercurrent of *Schadenfreude*. There was also a rueful note about the ravages of German hyperinflation. Meyerhof calculated, at the current exchange rate with the Swedish krona, that his share of the prize money would be 10^{13-14} marks; whereas, by the time he received it in Stockholm, it would be worth 10^{20} marks—a probable devaluation of one million times in six weeks.

Notes

1. C. Mulley, *The Woman Who Saved the Children: A Biography of Eglantyne Jebb, Founder of Save the Children* (Oxford: OneWorld, 2009).
2. R. Skidelsky, *John Maynard Keynes*, Vol. 1: *Hopes Betrayed, 1883–1920* (London: Penguin Books, 1994), 269.
3. Mulley (2009), 129.
4. C. Smith-Rosenberg, *Disorderly Conduct* (New York: Oxford University Press, 1986).
5. Mulley (2009), 131–132.
6. Mulley (2009), 131.
7. C. Angier, "Margaret Hill" (unpublished biography, vol. 1, 1978) 95–98, Churchill Archives Centre, Cambridge: The Papers of A.V. Hill (henceforth "AVHL") II 5/76.
8. F.A. Keynes to M.N. Keynes (1/9/12) AVHL II 5/117.
9. Mulley (2009), 132.
10. M.N. Keynes to E. Jebb (9/1/13, 12/1/13, 14/1/13) AVHL II 5/46.
11. M.N. Keynes to A.V. Hill (24/1/13) AVHL II 5/117.
12. M.N. Keynes to A.V. Hill (31/1/13) AVHL II 5/117.
13. Mulley (2009), 133.
14. Mulley (2009), 134.
15. M.N. Keynes to E. Jebb (6/2/13) AVHL II 5/46.
16. J.M. Keynes to D. Grant (13/2/13) in Skidelsky (1994), 269.
17. Mulley (2009), 134.
18. Mulley (2009), 140.
19. *The Times* (26/10/1923).
20. O.F. Meyerhof to A.V. Hill (29/10/23) AVHL I 3/58.

2
Not Even a Physiologist

It took Hill about twenty minutes to ride down from Highgate to Gower Street on Buster. There was a throng of excited students waiting to greet him on his arrival in the courtyard of UCL. They carried him on their shoulders around the College and then out onto Gower Street, halting the traffic and announcing to taxi drivers that their professor had won the Nobel Prize. Then they took him aloft to the attic laboratory of Ernest Starling, who was his predecessor in the Jodrell chair. At Hill's inaugural lecture, Starling had introduced him and joked that he did not know any physiology. The students now taunted him with that remark. Starling was unrepentant, exclaiming: "I said he didn't. He doesn't know a damned word!"[1]

Starling had determined that Hill should succeed him, after staying at his house in Manchester during a meeting of the Physiology Society in December 1922. He wrote to his daughter soon after, extolling Hill's qualities as "a tall vigorous personage, 35 years old, who will be the most important person in the physiology world—and likes to be in the scrum. A good fighter!"[2] Starling had been professor at UCL for over twenty years and, aside from his wide-ranging research work, he made considerable progress toward his dream of building an Institute of Medical Sciences there. The new physiology department, for which Starling largely raised the money, was opened as the first step in 1909. By 1922, the Rockefeller Foundation had committed to fund further stages. So, Starling was unabashed about telling the provost of UCL, Sir Gregory Foster, that his resignation to concentrate on his own research "is the crowning stroke in my creation or re-creation of a physiology school." Foster queried his choice of successor; Starling confirmed that Hill was the right man before adding, "But of course he is not a medical man, he is not even a physiologist." Foster gasped, "But he *is* a professor of physiology?" and looked "as though the bottom had dropped out of his Universe." Then Starling began to reveal the serious intent behind his levity. Predicting that Hill would do very well as head of the institute, which was already staffed with talented physiologists who were medically trained, Starling pointed out that it would be a great fit "to get a first-class physical man." He continued, "A man should always approach our goddess bearing gifts—it may be medicine—with its many sided training in science and mankind—or it may be physics or zoology—or chemistry."[3] The leading school of physiology in the country was at Cambridge (where Hill studied), but Starling thought that once Hill joined him and William Bayliss at UCL, "Cambridge will have to look to its laurels." Predicting that Hill would not return to Cambridge in the foreseeable future, he told his daughter that he wanted only "the best twenty years of his life for University College."

Bound by Muscle. Andrew Brown, Oxford University Press. © Oxford University Press 2023.
DOI: 10.1093/oso/9780197582633.003.0002

Hill always had a knack for impressing influential men after short acquaintance, and it had helped him at critical junctures throughout his youth. His father, Jonathan, ran a timber company that had been in business in Bristol since the mid-eighteenth century. His mother, Ada Priscilla, was also from Bristol, although from a poor family. Her grandparents in Exeter raised her to relieve the pressure on her parents. Ada recorded the birth of her first baby in her new *"Petite" Note Book* on a page about the size of a large postage stamp (Figure 2.1).

The young couple had been married for five years and lived in a substantial house with high sash windows, situated at the top of a street in Cotham that was lined with the homes of well-to-do merchants. Jonathan's younger sister and one housemaid, twenty-five years old (the same age as his wife), also lived in the house.

Ada, an excited young mother, recorded occasional landmarks in her little notebook. Her baby son was "talking plainly" at twenty months and, when barely two years old, delighted her by exclaiming: "Dear mother, dear father!" Ada gave birth to a daughter, Audrey Muriel, in February 1889, but within the year the happy family life had imploded. Ada discovered Jonathan to be having an affair with the nursemaid. She took her two small children and moved into another house owned by Jonathan in Cranbrook Road, Redland—half a mile from the original family home.

Ada must have been devastated by Jonathan's infidelity, although sexual liaisons between householders and their domestic staff were commonplace. Whether or not there was an attempt to repair the marriage, there was no reconciliation. Divorce was an expensive and complex business, with the law deliberately biased against women. Whereas a husband could obtain a divorce on the grounds of his wife's adultery, a wife had to show adultery plus some added transgression, such as desertion or cruelty. So immediate divorce was not an avenue open to Ada, and she made the decision to devote her life to rearing her two children. She did not talk to them about their father.

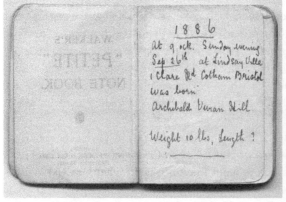

Figure 2.1 Ada Priscilla Hill with her baby Archibald Vivian and her record of his birth
Photograph courtesy of Professor Nick Humphrey/Churchill Archive Centre, Cambridge (Papers of A.V. Hill, AVHL).

Hill was burdened with the names Archibald Vivian, which he would do his best to avoid in later life. His mother called him "Vivian," and her love for him was reinforced by great determination. She taught him the three "R's" at a young age, and he quickly exhibited a facility for learning. He also made his first foray into the world of physiological experimentation. His grandparents kept chickens, and Vivian, a curious and impatient little boy, thought their eggs were taking too long to hatch. So, he cracked some open to see what was happening. He then decided to sit on the remainder.[4]

In the summer of 1894, Ada moved with her two children to Weston-Super-Mare, a seaside resort on the muddy Somerset coast about twenty miles southwest of Bristol. By scrimping and with some support from relatives, Ada managed to save enough money to send Vivian to Brean Villa School, a well-established private school in the town. He started there just before his eighth birthday and settled in well. Classes were small, and Vivian soon shook off his initial shyness. Reports noted his good memory and by Easter 1895 he was top of his class.[5] He was competitive on the games field, especially playing hockey, and noted to be a bad loser. He also began to enjoy long solitary walks and hiking up hills in the surrounding countryside. Ada, who was ambitious for her son, decided that he should assume the role of a young gentleman. On Sundays she dressed him in the style of a young toff: an Eton suit, a white shirt with a starched collar, and a top hat. The street urchins of Weston asked whether he kept rabbits in the hat. Others would entertain themselves by throwing stones at him in the street, which he soon learned to return with vigor and accuracy.

Vivian's progress at Brean Villa was so impressive that at the end of his second year, he was to be entered for the College of Preceptors examinations to qualify for a secondary school. This premature plan was voided when he contracted mumps and was confined to home for several weeks. He was visited by a family physician, Dr. Fligg, who began to take an interest in the clever boy and became the first to decide to support him. Vivian passed the College of Preceptors examination in 1897 but stayed at Brean Villa until 1900, when he was top in every subject. He was reported to be gaining more control over his temper, although still needing to show better grace when defeated on the sports field. While he had some scholarship monies, his mother still had to find nearly ten pounds per term to cover tuition fees, school uniform, and materials for the classroom. Dr. Fligg recommended to Ada that she should move to Tiverton in Devon so that Vivian could attend Blundell's School. Generously, he offered to help her with the fees. Devon was Ada's spiritual home—her speech retained a soft West Country burr—and for the first time since her marriage, she had a pledge of financial support.

Tiverton stands on the Rive Exe about ten miles north of Exeter. It was first settled in Saxon times and is surrounded by small, sleepy Devon villages. Tiverton's past, however, is a lively microcosm of English history: entries in Domesday Book, when William the Conqueror was the major landowner; a Norman castle; the Black Death and other plagues; riches from the wool trade in the sixteenth and seventeenth centuries; a royalist stronghold during the Civil War (when the castle was besieged by Roundheads); several devastating fires; canals, railways, and industrialization. Peter Blundell was one of the wealthiest wool merchants from Tiverton in the sixteenth century. He never married and on his death in 1601, he left a bequest to found a free grammar school, as well as to fund six closed scholarships to Oxford and Cambridge for boys from the school to become puritan preachers.[6] In the late nineteenth century, the school moved to the outskirts of Tiverton into attractive new buildings, constructed from the local red stone.

Ada and her two children arrived in Tiverton in the spring of 1900. There were between 250 and 300 boys in the school under the direction of A.L. Francis, the headmaster. A Cambridge M.A., Francis was the first headmaster in the history of the school not to be an ordained clergyman. Although Peter Blundell clearly anticipated that his school would produce men of the cloth, he did not stipulate in his will that the headmaster must be ordained. Francis's appointment was an example of the diminishing influence of the Church of England in educational establishments in the Victorian period. The headmaster and other members of staff contributed generously to the costs of new school buildings that opened in 1882.

Ada managed to rent a small, terraced house within an easy walk of the new Blundell's School, and Vivian entered as a day-boy. By the end of his first term, his schoolwork was already judged to be "brilliant, full of intelligence and energy."[7] But for his poor showing in ancient Greek, he would have been top of his class. In June 1901, Hill took scholarship exams at Blundell's. There was quite a heavy emphasis on mathematics that played to Hill's strong suit. One question: *"A man can row 15 miles upstream and back again in 6 hours; and he notices that it takes as long to row 3 miles upstream as 5 down. What was the rate of the stream?"** This was one of ten questions to be tackled in a two-hour algebra paper. Apart from the mathematics, there were papers in French, Latin, and history. During the school holidays, Hill was expected to read and commit to memory great chunks of Shakespeare. He also studied the Bible with pleasure and was inevitably introduced to R.D. Blackmore's romantic novel, *Lorna Doone*. Blackmore had been a student at the old Blundell's, six decades before. The adolescent Vivian may well have identified with John Ridd—the action-man hero of the story who ends up with Lorna, despite seemingly impossible odds—even though Ridd fights a day-boy on the school grounds at one point. Blackmore's lyrical descriptions of the Devon and Somerset countryside, with the hills, moors, streams, and crags, ensured that *Lorna Doone* remained Hill's favorite novel.

---

* Let x be the rate of the stream (mph) and y the rower's rate (mph)
Time to travel 15 miles upstream = $15/(y-x)$ and back downstream $15/(y+x)$
$15/(y-x) + 15/(y+x) = 6$
We are told that $3/(x-y) = 5/(x+y)$ and by cross-multiplying, $y=4x$.
Rewriting: $15/(4x-x)+15/(4x+x) = 6$
Solving: $6x=8$ so $x = 1.33$ mph

After securing the Foundation scholarship, Hill took his place in the top mathematics class taught by J.M. "Joey" Thornton, a down-to-earth Yorkshireman. Thornton had placed fourth in the competitive Mathematics Tripos examination at Cambridge University in 1882, and on graduation became a master at Blundell's. Thornton was immediately impressed by the ability of his new student: "I like his work very much," he reported to Mrs. Hill. For her son, Thornton was simply the best teacher he ever encountered. While allowing that not all his students could be brilliant, Thornton conveyed the notions that "carelessness, looseness and inexactness in thought, speech or habit are not only foolish and unprofitable but despicable and wrong."[8] Joey Thornton was not immune to these foibles himself outside the classroom. He infamously built a garage adjacent to his house, where the dimensions corresponded so tightly to those of his car that he was unable to open the driver's door.

Vivian's successes were not restricted to the classroom. He would represent the school in team events: a mainstay of the Hockey 1st XI for all four years, he also played cricket and rugby for the day-boys in house matches. As a tall, long-legged youth, he became an enthusiastic athlete and dominated the 1902 Sports Day winning the under-16 100-yard dash, the quarter-mile and mile races, as well as the long jump. At his final Sports Day at the school in 1905, he won the quarter-mile in a time of 59 seconds and the mile in 5 minutes 3.2 seconds. Although new to rifle shooting, he was talented enough to win the national under-16 competition at Bisley in 1902 (Figure 2.2).

He continued to excel in scholastic achievements: he won the school's main mathematics prize in his final two years. In June 1904 he obtained a Huish Exhibition, open to boys from Blundell's and other West Country schools, which guaranteed him an income of 50 pounds per year if he went to Cambridge or Oxford Universities. A scripture paper was a requirement of the Huish examination so that Vivian stepped up his Bible study. This became a lifelong habit, but not one based on religious faith. Blundell's had a long-standing relationship with Balliol College, Oxford, but as a budding mathematician Hill's ambition was to go to Cambridge. Sidney Sussex was the Cambridge college linked with Blundell's, but Hill aimed for the apex of British mathematics: Trinity College. He went to Cambridge to take the entrance examination, knowing that without an additional scholarship the 50 pounds per year would not be enough for him to go to the university: yet he remained calm and said he was less nervous than before a mile race or speaking at the Debating Society at Blundell's.[9] His confidence and ability brought him the top Open Scholarship in mathematics to Trinity College.

Hill emerged from Blundell's a happy, energetic, tall, young man with the self-confidence that comes from triumphs in the classroom and on the sports field. For a boy from a broken home, his progress was quite remarkable. He regarded Blundell's as "the moving staircase" that elevated him to the prospect of a fulfilling life but was in no doubt that his mother's love and devotion for her children was the essential foundation. He admired his mother and never seemed to display the irritation or

Figure 2.2 Hill with sports trophies, Blundell's School, 1905
Photograph courtesy of Professor Nick Humphrey/Churchill Archive Centre, Cambridge (Papers of A.V. Hill, AVHL).

embarrassment of an adolescent son: they shared equitable temperaments. Living at home, he saw more of his mother than he would have done as a boarder; her constant presence probably meant that he worked harder than his friends, but he was by no means a narrow swot. His sister, Muriel, was also a good student and not overawed by her older brother. Ada would stay in Tiverton until Muriel finished her schooling. Although Vivian won two scholarships to Cambridge, Ada was still worried about money and wrote to W.W. Rouse Ball, senior tutor and director of mathematical studies at Trinity College Cambridge (TCC), seeking advice on minimizing living costs. Rouse Ball's reply was terse:

> Dear Madam,
> Of course, I can offer your son only such rooms as happen to be available—but I note that you would wish them as cheap as possible. I believe he would find it rather cheaper to go first into lodgings and if he likes to move later into College. Furnished lodging rooms can be got at a price of six pounds ten shillings to seven pounds per term.
> Yours truly,
> W. W. Rouse Ball

A second letter followed in July 1905 from Rouse Ball, assigning A.V. Hill a set of rooms, L5, in the Great Court of Trinity with a rent of five pounds per term plus a service charge of two pounds eight shillings.[10] In early September, a more welcoming letter followed from a newly appointed tutor at Trinity, Walter Fletcher, who was expecting his first crop of undergraduates. He advised Hill to send his bed linen in advance to the Trinity bed makers and that he would call to see him at L5.[11]

One of Hill's favorite walks in Tiverton was to the top of Exeter Hill from where there is a panoramic view that on a clear day can reach Dartmoor. To celebrate his scholastic success, Ada arranged to take her two children to a farmhouse on Dartmoor during August 1905. Hill found it to be a magical place, where he could climb Cawsand Beacon or run for miles across open country. The farmer lent him a gun and he went out hunting for rabbits. Early one afternoon, he noticed an eclipse of the sun developing. He opened his pocket watch and smeared the glass with the warm, sticky blood of a rabbit so that he could view the celestial event safely. The eclipse was almost total: the date was 30 August.[12]

One month later, it was time for Ada to take her son up to Cambridge. His diary[13] records that they arrived on 29 September and called on Mr. Fletcher. Ada must have again brought up her anxiety about money. Days after meeting the Hills, Fletcher wrote to the headmaster of Blundell's, A.L. Francis, pointing out that Hill had already received from TCC "the highest emolument the college can offer."[14] As his tutor, Fletcher generously offered to help out "from the very limited funds at my disposal," but during the first term, while Hill was proving himself, the Governors of Blundell's would be approached for extra financing if necessary. At their autumn meeting in 1905, the Governors awarded Hill an immediate grant of thirty pounds.

Hill can hardly fail to have been impressed by the grandeur of Trinity College. Founded by Henry VIII, the College was mostly laid out by its Master, Thomas Nevile, at the beginning of the seventeenth century. One enters through the Great Gate and instantly leaves behind the bustle of Trinity Street for the quiet splendor of the Great Court. To the right in the dark, northeast corner of the court, Isaac Newton had lived on E staircase. L is in the southern wall of Great Court that contains the stunning Queen's Gate; Hill's set of two rooms was at the top of staircase L. On his first night,

Hill had two visitors: one from the Cambridge Intercollegiate Christian Union and one from an atheist. He was not enticed by either, although he took a personal dislike to the Christian who predicted that he, not Hill, was going to be the top mathematics student or "Senior Wrangler" in 1907.[15] Hill's matriculation at TCC marked the end of the name Vivian, by which he was known at Blundell's. Archibald had always been kept under wraps, and from now on, he would just be known as "AV."

Hill jotted down in his diary the lecture schedule for the Michaelmas Term. He had just one hour of lectures every morning including Saturdays, and with afternoons free. The Reverend Ernest Barnes, a pure mathematics don in his early thirties at the peak of his powers, lectured on differential equations and was also Hill's director of studies at TCC. They would have weekly tutorials for which Hill would write essays, and a close bond soon developed between the pair. Hill's personal tutor, Walter Fletcher, a young physiologist, shared his love for running—he had been president of the University Athletic Club in 1888. Alfred North Whitehead, who taught analytic statistics, was older but he and his wife also showed AV consistent kindness. In addition to giving occasional lectures, Whitehead was deeply occupied in writing *Principia Mathematica* with his former student, Bertrand Russell. Hill recognized that Whitehead's immersion in the philosophy and structure of mathematics was beyond his grasp. He described Mr. Whitehead talking "funnily about time and space." In a later term Whitehead "appeared to deduce the principles of geometrical optics by drawing lines on a blackboard." Hill admitted that "these lectures meant nothing to me; I never was an intellectual."[16]

While he may have lacked the imagination and insight of Whitehead or Russell, Hill was a strong mathematician and, importantly, a competitive student. In the ever-present quest for financial support, he entered for another major Trinity College scholarship at the end of his second term. There was a three-hour mathematics paper with a dozen advanced calculus, trigonometry, and algebra problems to be solved. The next day was a general paper where six to eight essays were to be attempted. Two specimen questions: "*Define and illustrate: mannerism, hyperbole, bathos and Irish bulls*" and "*Discuss the origin of the English thoroughbred horse.*" Hill's award of the scholarship brought warm letters of congratulation from Fletcher and from the Master of Trinity, Henry Montagu Butler, who was in the traditional mold for an Oxbridge head of house—a classicist and cleric.

For most of the nineteenth century, the teaching of mathematics at Cambridge University was dominated by the Tripos examinations. These were taken in two parts—the first associated with public notice and the second involving a deeper exploration of selected topics. The first-part exams especially were a punishing test of mental and physical stamina: two papers per day over two weeks or so, and hundreds of questions that needed to be answered quickly to rack up a good score. The questions were tricky as well as technically demanding. Dons would often submit their recent research work in the form of an examination problem, to see how fevered young brains might treat it. Neither poring over textbooks nor attending lectures from college dons would impart to the Tripos candidate the wide range of topics he was going to encounter, or the techniques needed to solve them. Students who wished to make their name in this frenzied contest would place themselves in the hands of private coaches. The top-scoring candidate was known as the Senior Wrangler, followed by second, third, and fourth Wranglers down to the Wooden Spoon, the title bestowed on the lowest-scoring honors student. Results were listed in the Cambridge and national press, and photographs of the Senior Wrangler appeared on postcards. Their coaches were as well publicized as racehorse trainers winning the Derby.

George Darwin, a fellow of Trinity and son of the naturalist Charles Darwin, criticized the Mathematical Tripos in his inaugural lecture as the Plumian Professor of Astronomy in 1883. Although he had been Second Wrangler himself, Darwin worried that the system did not prepare young mathematicians for a research career and instead gave the impression that "if a problem cannot be solved in a few hours, it cannot be solved at all." It was his experience that the nub of a research inquiry might reveal itself only after months or years of attack, during which time the initial "chaos of possible problems" had to be "pared down until its characteristics [became] those of a Tripos question."[17] Defenders of the traditional Tripos countered that the emphasis on problem solving meant that the most able students developed remarkable fluency in their selection and application of analytical mathematical methods, especially in the field of physics. Lord Rayleigh, Senior Wrangler in 1865 and the first Nobel laureate from Trinity College in 1904, was a master of mathematical physics who relied on physical intuition combined with a personal armamentarium of tested mathematical methods. Rayleigh believed that an overexposure to pure mathematics could be detrimental to the mathematical physicist.

By the early years of the twentieth century, the pendulum was swinging against Lord Rayleigh and toward Sir George Darwin's side. Trinity College boasted the greatest array of young pure mathematicians in England; they wanted their specialty to separate itself from the broader mixed mathematics that had prevailed at Cambridge for more than a century. They believed that the structure of the Tripos examinations was detrimental to the analytical and imaginative insights required for the abstract nature of their subject. Undergraduates too were becoming unwilling fodder for a process that could see them labeled as failures, if they did not make it into the upper echelons as a Wrangler. With the syllabus naturally expanding over the years, the most able students differed in which subjects they favored, so that it was becoming impossible to rate their merits on the basis of a common exam.

The Trinity don in the ascendancy was G.H. Hardy—a handsome, nimble, sardonic man, overflowing with aphorisms and opinions. He was the secretary of the committee that persuaded a reluctant University Senate in 1907 to abandon the order of merit for Wranglers: he had been Fourth Wrangler himself in 1898 and so did not do so, he said, as "a mere jealous outsider." But the new system would not be introduced until 1909, which meant Hill was tested under the traditional, pressured contest. Having secured extra scholarship funding, he could afford to engage the last of the great coaches, Robert A. Herman, a Trinity don with a "genius for teaching"[18] and a fine record of producing winners. There were ten mathematics Tripos papers, each of three hours' duration, tabled from mid-May to 1 June 1907. Most of the hundred or so questions involve symbols and equations comprehensible only to advanced mathematicians. As a measure of their abstruseness, consider that the introductory part of one complex question asked the candidates to: "*Investigate the differential equations of equilibrium of a flexible, inextensible string which is acted on by a system of forces in one plane.*"[19]

The results for the Tripos were posted on the door of the Senate House on Tuesday, 11 June. His close friend, Charles G. Darwin,[20] Sir George's son, accompanied an excited AV to await the results. Darwin, a large, cheerful, energetic youth, who did not seem to take life very seriously, was one year behind Hill at Trinity College. When the list appeared, Hill's was the name in third place. He was "badly pummelled by Darwin" in celebration; another Trinity friend, the Orientalist Harry St. John Philby, joined in.[21] Trinity students had a banner year, filling the top five spots. Despite his high finish, AV might have been chagrined to see the name of George Watson, the proselytizing Christian, as Senior Wrangler. AV sent a telegram to his mother before enjoying a huge breakfast with Darwin and wandering around the backs next to the river in a daze. Several newspaper reporters interviewed him, and he had his photograph taken. He went to see his supervisor, Barnes, with whom he decided to study for the Part II of his degree the following year. Rouse Ball wrote to him that day offering "hearty congratulations" but hoping that he was not disappointed with the position of Third Wrangler and did not regret taking the Tripos in his second year.[22]

Despite his tremendous success, AV demonstrated the limitations of the traditional contest in selecting students for the highest level of research. G.H. Hardy regarded mathematics as "a young man's game"—one based on intellectual curiosity and requiring professional pride as well as a degree of immodesty to be a successful practitioner of the art.[23] Hardy drew a distinction between *real* mathematics that has an inner significance and beauty, that is incomprehensible to all but a few academic experts, and everyday or trivial computation. AV undoubtedly had a facility with numbers: he delighted in using them and understanding their intricacies, but, as he knew, he could not be confident of finding his way in the most arcane corners of the mathematical world.

A crucial letter came from R.A. Herman, who coached the top three Wranglers in 1907:

> Dear Hill,
> I think you would be well advised to give up Part II and concentrate on the Civil Service. For that you will have ample time to be well prepared; and with your attitude to mathematics, I do not think the prospects of a Fellowship are particularly bright, and there may be much time wasted. Your abilities are enough for Part II, but original research, especially in applied, is a very quagmire and then what to get to do is the rub.[24]

Reading this unvarnished, unsolicited advice must have been a shock after weeks of congratulations from friends, family, and indeed from others at Trinity. Yet it confirmed a gnawing doubt that Hill was already experiencing. His mother would be uncomprehending after her son had just triumphed in the most famous university exams in the country. AV placed great trust in his personal tutor, Walter Fletcher (Figure 2.3). Previously he had discussed with him his uneasiness in persisting with mathematics. Fletcher had asked him then why he did not switch to physiology.

Confronted by Herman's cautionary letter, Hill now contacted Fletcher for urgent advice. He responded, leaving Hill to make the choice that suited his circumstances. He advised him to switch to the Natural Sciences Tripos, taking physics and chemistry. For the third required subject, Hill should take geology if he wished to graduate in one more year. If he was prepared to take two years, he should pick physiology, but his grounding in biology was not strong enough to complete it in one year.[25] The Mathematics Tripos contained elements of physics, such as electricity and optics, but no laboratory work. Often students, though not the top handful of Wranglers, decided to transfer to the natural sciences after completing the first part of the mathematics degree.

Hill quickly made up his mind that he would spend two more years to graduate and take physiology. His sister, Muriel, wrote to her brother from school, saying that

Figure 2.3 Walter Morley Fletcher (1873–1933)
Photograph courtesy of the Wellcome Collection, London.

"Mr Chalk sends his love and is glad you have given up the dry bones of mathematics and are doing something fairly sensible."[26] She also included a ribald account of her German lessons to amuse, and perhaps shock, her straitlaced big brother:

> Mr Selopp has progressed from asking me to translate: "Let us love one another" to "We love one another" to "I love you and you love me," "Thou art mine and I am thine." People are beginning to be suspicious. Last Monday he suddenly asked: "*Fräulein* Hill decline my trousers" causing a perfect shout of laughter.

Over the summer, Hill bought science books and made a start on chemistry. In mid-August, he took a train journey to Scotland, where he met up with Cambridge friends for one month of hiking, shooting, fishing, and biking. He returned to Devon in time for his twenty-first birthday, before going back up to Cambridge in October.

Physics was in the midst of revolutionary changes in 1907. The Cavendish Laboratory, under the direction of J.J. Thomson, was one of the most exciting research centers in the world. Ten years before, Thomson measured the electric charge-to-mass ratio of the stream of fast-moving particles that constitute cathode rays in a vacuum tube. In 1899, he was able to prove the mass of these particles was thousands of times smaller than that of atoms, previously believed to be indivisible. He named the subatomic particles *electrons*. In 1906, Thomson emulated Lord Rayleigh, his immediate predecessor as Cavendish professor, by winning the Nobel Prize for physics. The discovery of the first subatomic particle naturally led to the hypothesis that electrons are the fundamental building blocks of matter. Thomson came to the view that an atom contains thousands of negatively charged electrons, balanced by some nebulous positive electrical charge (needed to render the atom electrically neutral), in what became known as his "plum pudding model" of the atom. Attracted by Thomson's achievements, the numbers of students coming to do research work at the Cavendish from elsewhere in the country, and indeed from around the world, had shown a steady increase since the turn of the century.[27] X rays and radioactivity were other recently discovered phenomena that would provide valuable clues to the structure of atoms. Finally, the classical theoretical basis of physics, dating back 250 years to Newton, was being challenged with the introduction of quantum theory by Max Planck, Albert Einstein, and their collaborators in continental Europe.

The physiology department at Cambridge was essentially created by the unstinting efforts of one man, Michael Foster. He was medically qualified and held the chair in physiology at the Royal Institution before agreeing to move to Cambridge in 1870.[28] Starting in one room with three tables and a few microscopes, he built up the most active group of physiology researchers in the country. He claimed a broad swathe of territory for his discipline, including physiological chemistry (biochemistry), histology (the microscopic study of biological tissue), and embryology.[29] Foster was not an inspirational lecturer, but he regarded chalk and talk as inferior to practical classes, where students observed and manipulated the activities of biological systems. He also founded the *Journal of Physiology*, wrote a popular textbook, and was the instigator of the International Congress of Physiology that first assembled in 1889.

Foster's choice to succeed him as head of the physiology department was J.N. Langley (Figure 2.4), who had worked in the laboratory since graduating in the Natural Sciences Tripos in 1874. Like Hill, Langley was the product of a school in Devon and spent his first two years at Cambridge reading mathematics. His main research work was on the autonomic nervous system that controls many of the body's involuntary functions: Langley introduced the terms "sympathetic" and "parasympathetic" nerves.[30] He was the joint editor of the *Journal of Physiology* with Foster, until the journal was foundering in 1894, when he paid off its debts, took over as sole editor, and enforced a more disciplined style.

Hill instantly felt more at ease studying empirical natural sciences than he had been in the more abstract world of mathematics. At the end of the year, he took the Part I of the Natural Sciences Tripos and still seemed to be harboring some doubts about his future. Butler, the Master of Trinity, offered to meet him, "to see your way a little clearer."[31] When the results of the examinations became available, he was able

Figure 2.4 John Newport Langley (1852–1925)
Photograph courtesy of Professor Nick Humphrey/Churchill Archive Centre, Cambridge (Papers of A.V. Hill, AVHL).

to write again to inform Hill that he had been awarded the Past Scholars' Prize.[32] He celebrated by buying a bicycle and rode it down to Devon for the summer vacation. As expected, he sailed through the Part II Tripos exam in the summer of 1909 with first-class honors in physiology. In doing so, he secured the G.H. Lewes scholarship that would allow him to begin research in the physiology department. His mother would again be close at hand. Muriel was accepted at Newnham College to read for the Natural Sciences Tripos, although as a woman she would not be eligible for a Cambridge degree. Ada left Tiverton to join her two children in Cambridge.

When Hill arrived to start his research career, his assigned bench was in a poorly drained basement, surrounded by cages of laboratory rats. There was another work-space partitioned off behind Hill's apparatus and the rats. Keith Lucas, who worked there, would often stop by to discuss research problems with him, as did Edgar Adrian who joined the laboratory in 1911. Hill formed the impression that "there were probably more great physiologists there to the square yard than in any other place, before or since; and not only because there were so few square yards."[33]

Foster's successor, J.N. Langley, showed an undiminished appetite for original research despite the manifold tasks of overseeing the burgeoning laboratory being added to his role as the sole editor of the *Journal of Physiology*, not to mention becoming a husband and father for the first time in his fifties. A series of experiments to elucidate the workings of the sympathetic nervous system (responsible for "flight or fight" reactions) came to an end in 1905. For this research, he employed the drug nicotine to stimulate the sympathetic nerve cells, and in so doing greatly clarified their anatomic arrangement and function. Langley decided to concentrate next on the junction between larger nerves and skeletal muscles in vertebrate animals.[34] Using anaesthetized fowl for his experiments, he injected drugs via the jugular vein, while observing the duration and extent of contractions in the leg muscles. Injection of nicotine in small quantities caused prolonged contractions: this effect was blocked by injection of a sufficient dose of the paralyzing poison known as curare. If he cut the sciatic nerves to the muscles, the effects of nicotine and its antagonist, curare, were essentially the same, even weeks later, leading Langley to conclude that the drugs act on the muscle tissue and not upon the nerve endings. The possibility of nerve regeneration was ruled out by electrical stimulation of the severed end, and degeneration of the nerve endings confirmed by microscopic examination. Langley further deduced that the effects did not depend directly on the contracting muscle fibers themselves.

> Since, in the normal state, both nicotine and curari [sic] abolish the effect of nerve stimulation, but do not prevent contraction from being obtained by direct stimulation of the muscle or by further adequate injection of nicotine, it may be inferred that neither the poisons nor the nerve impulse act directly on the contractile substance of the muscle but on some accessory substance.
>
> Since this accessory substance is the recipient of stimuli which it transfers to the contractile material, we may speak of it as the *receptive substance* of the muscle.[35]

Langley suggested that many plant-derived alkaloids in addition to nicotine as well as "other internally secreted substances" (hormones) act in the same way. He distinguished between two constituents within cells: one concerned with the chief function of the cell, such as contraction in the case of muscle cells or secretion for glandular cells, plus "receptive substances which are acted upon by chemical bodies and in certain cases by nervous stimuli." Over the next few years, he lectured widely on his new ideas in England and in European centers. As one might expect, some scientists were skeptical about his somewhat speculative chemical theory of drug action.[36] His critics tended to favor a process that depended on a concentration gradient between the outside and the inside of the cells in the target tissue, rather than one that was triggered by the chemical binding of the drug to a receptor substance.

Langley normally took new research scholars under his wing when they arrived in the laboratory. So it was with A.V. Hill in the autumn of 1909. He was soon aiding other, less numerate, physiologists in analyzing their results. Langley wanted to see if confirmatory evidence for his chemical theory could be generated. If the effects of drugs such as nicotine and curare were due to chemical reactions with unidentified receptor substances in muscle cells, as Langley believed, could the energetics of those reactions be uncovered?

It was a fiendishly difficult problem to set a neophyte researcher, but perhaps Langley believed that Hill's mathematical prowess would allow him to reach a fruitful conclusion. Hill assembled bits and pieces of apparatus from Langley himself and from Keith Lucas, who had designed a recording system so that muscle contractions could be traced onto a slowly rotating drum.

Leon Orbeli, a Russian physiologist and acolyte of Ivan Pavlov's, arrived from St. Petersburg to gain experience in the leading British laboratory, just at the time Hill was beginning his work.[37] Orbeli was disarmed by the courtesy and ease of Cambridge academics, and he recorded his impressions of various figures including Hill. Langley suggested that Orbeli should study the influence of sympathetic nerves on the blood vessels of frogs; until he had obtained the necessary animal license, he was allowed to work only on dead frogs. When Orbeli began operating on live ones, a man with a pipe approached him and, introducing himself as Hill, asked why the Russian scientist was sterilizing his instruments. A somewhat surprised Orbeli explained that microbes from the skin could otherwise cause infection, whereupon Hill claimed ignorance of bacteriology or medicine! Then Hill "disappeared somewhere" for a week or so, "appearing in the laboratory to smoke his pipe and drink a cup of tea." Tea was served every afternoon at 5 P.M. in the departmental library and "suddenly everyone appeared from whatever they were doing, reading or working, and for thirty or forty minutes the whole laboratory drank tea and exchanged the latest news. Then everyone disappeared into his own corner and got back to work." At various times, Orbeli inquired about Hill and was told that he was spending his time reading, or that "he is at an instrument factory watching the making of his galvanometer." When Hill took delivery of the galvanometer (an instrument to detect tiny electric currents), he set it up on a special mounting in the basement, where according to Orbeli, "he shut himself up and nobody saw him for a week." Then "suddenly he finished, came out again," resumed running, "and all sorts of fun started up again."

Although the exact duration of Hill's experiment is unknown, one week is probably a reasonable, if astounding, estimate. What we do know is that Hill's first scientific paper was published in the *Journal of Physiology* just before Christmas. Even though the journal was Langley's personal title and he had most to gain from Hill's work, he always read papers scrupulously and was famous for requiring extensive rewrites. And there must have been a hiatus between assembling the December issue and having it printed.[38]

Hill observed the time-course of the contractions in the *rectus abdominis* muscle of the frog while immersed in solutions of nicotine of different concentration. The nicotine solution was then washed off the muscle with Ringer's physiological solution, and the muscle specimens would relax completely over time. He found that both the contraction and relaxation phases could be accurately represented by two equations, both simple exponential functions. He then posed the question as to whether these curves could be due to a physical process, where the degree of contraction is due to nicotine passively diffusing into the muscle fibers, or whether his experimental results comported with Langley's hypothesis that the nicotine undergoes a chemical reaction with a receptor substance. He made the important point that such a chemical reaction must be reversible since the muscle relaxes completely in Ringer's solution (a mixture of salts dissolved in water to replicate serum, minus the protein). His results from about twenty different curves showed a consistent hyperbolic shape that could be described by a formula for the degree of contraction that Hill argued was exactly of the form required and "is very strong evidence in favor of the hypothesis of a combination between nicotine and some constituent of the muscle." He further strengthened the case by studying the action of curare as an antagonist of nicotine and by comparing the contraction curves of the two halves of the abdominal muscle immersed in nicotine solutions at different temperatures. It was compelling evidence that there was a reversible chemical combination between nicotine or curare and a receptor substance in muscle. In addition to thanking Prof. Langley for his kind help, Hill acknowledged Mr. C.G. Darwin of Trinity College in the journal paper for "his aid in my analysis of the curves."

Sixty years later, Hill was overmodest about the significance of this first publication, saying it was undertaken to satisfy Langley and that "the results look quite unimportant today." But when his distinguished colleague and friend Bernard Katz read the paper closely, while writing Hill's obituary for the Royal Society,[39] he identified two important features buried in the main mathematical argument. It was, Katz pointed out, the first kinetic description of drug-receptor interaction. Hill's mathematical treatment also foreshadowed the Michaelis-Menten equation of 1913 that is perhaps the most famous in all of biochemistry, dealing with the reversible reactions of enzymes on substrates. Katz also showed that Hill had exactly anticipated the 1918 equation of Irving Langmuir that deals with the adsorption of gases on metal surfaces. His reassessment of the significance of Hill's first publication has gained general acceptance by modern pharmacologists and physiologists. A modern-day systems biologist at Harvard University remarked: "Why his first published paper faded from sight remains a mystery, but it remains the seed from which quantitative pharmacology subsequently flowered."[40] Another review paper published in 2006 begins by stating: "Pharmacology started to develop into a real quantitative science in 1909, when A.V. Hill derived the Langmuir equation in the course of his studies on nicotine and curare."[41] In a similar vein, Hill is now recognized as one of the founding fathers of receptor theory, "pharmacology's big idea,"[42] at last receiving the credit he deserves by being memorialized in the Hill-Langmuir equation. By the time Hill died in 1977, the hypothesis that drugs have their effect by targeting receptor proteins on cell membranes was well accepted, but those receptors were not widely characterized at a molecular level until the last two decades of the twentieth century.

In addition to the experiment for Professor Langley, Hill had been collaborating with the senior demonstrator in the laboratory, Joseph Barcroft (Figure 2.5). Barcroft, an Ulsterman, was an acknowledged expert on the uptake of oxygen by different organs and their output of carbon dioxide. His improved methods for determining the concentrations of gases in the blood led him to study the function of

Figure 2.5 Joseph Barcroft (1872–1947)

Photograph courtesy of Professor Nick Humphrey/Churchill Archive Centre, Cambridge (Papers of A.V. Hill, AVHL).

hemoglobin—the molecule that carries oxygen from the lungs to the tissues. It was generally accepted that oxygen molecules formed a reversible chemical bond with hemoglobin molecules, as governed by the law of mass action, but a young chemist in Germany, Wolfgang Ostwald, disputed this.[43] He regarded hemoglobin as a colloid or mixture of particles, rather than a true molecule, and proposed that oxygen was held in combination by physical adsorption (a mechanism not clearly defined). The dispute was akin to the argument raised against Langley's receptor substance theory, so it was unsurprising that Barcroft would discuss the problem with Hill.

The pair devised an experiment to gather fresh evidence to support the notion that the reversible combination of oxygen with hemoglobin was indeed a chemical process. Barcroft designed the experimental setup and made the measurements of oxygen content in the blood samples; Hill's responsibility was the mathematical analysis of the observed data. Barcroft dialyzed the blood sample to remove any salts from solution and started with oxyhemoglobin (fully saturated with oxygen). He then plotted the time-course of the oxygen saturation decreasing as he bubbled nitrogen gas through the solution. He found that when the process was carried out at a constant temperature of 38° C the hemoglobin desaturated much faster than it did at 18° C. At the end of his complex analysis, Hill was able to conclude that "the velocity of dissociation of oxyhemoglobin obeys an equation derived from the laws of mass action and has a high temperature coefficient." Having satisfied themselves that the reversible union of oxygen with hemoglobin is indeed a chemical reaction, Barcroft and Hill set about to determine the molecular weight of hemoglobin. On Hill's part, this involved calculating, from the second law of thermodynamics, the heat that would be released from combination of one gram-molecule of hemoglobin with oxygen. Barcroft bubbled oxygen through a solution of reduced hemoglobin in a water bath and measured tiny temperature increases of the order of 0.001° C. The ratio of Hill's calculation and Barcroft's measurement of the heat released by 1 gram of hemoglobin yielded a minimum possible molecular weight for hemoglobin of 16,669.

Hill's joint paper with Barcroft appeared in the same volume of the *Journal of Physiology* as his study of nicotine and curare on muscle contractions, serving notice that another industrious physiologist with a rare mathematical ability was making his mark in Cambridge. His next paper, also on hemoglobin, was published the following month, after it had been presented at a meeting of the Physiological Society. In 1904, a Danish physiologist, Christian Bohr, and his colleagues showed that the percentage saturation of hemoglobin with oxygen plotted against the partial pressure of oxygen gave a sigmoid or S-shaped curve.[44]

If the concentration of carbon dioxide, CO_2, in the blood increases, the oxygen dissociation curve shifts to the right ("the Bohr effect"). The implication of the sigmoid curve is that at low oxygen levels, hemoglobin does not readily take up oxygen so that more is made available to the tissues. The partial pressure of oxygen in venous blood is about 40 mmHg, at which level about 30 percent of oxygen is unloaded; at arterial levels of 96–98 mmHg, the hemoglobin molecule is almost completely saturated with oxygen as bright, scarlet-colored oxyhemoglobin. The shift of the Bohr effect to the right serves the needs of the organism in that more unbound oxygen gas is made available to organs in the presence of higher carbon dioxide levels.

Until Hill gave his talk, no one had given a good explanation of why hemoglobin was reluctant to bind oxygen at low pressures, while its affinity for the gas rises the more oxygen is present. Hill suggested it could be "due to an aggregation of the hemoglobin molecules by the salts present in the solution."[45] He and Barcroft used hemoglobin that had been dialyzed to remove all extraneous salts, while Bohr in his original experiment had used whole blood. For their dialyzed hemoglobin, Hill and Barcroft were sure that each molecule contained just one atom of iron. Others had proposed higher molecular weights (indeed some chemists still believed proteins were shapeless colloids of infinite molecular weight): Hill suggested that two or three or more single molecules of hemoglobin could stick together or aggregate.

Hill came to think that the S-shape of the dissociation curve meant that the fractional saturation of hemoglobin would be proportional not to the partial pressure of oxygen, but to a power function of the pressure, giving a sigmoid curve. He devised a fairly simple equation to describe the process by which the more oxygen is bound by hemoglobin, the easier it becomes to bind more.

$$y_1 = 100 \ (Kx^n/1 + Kx^n)$$

Where, y_1 = the percentage of hemoglobin saturated with oxygen

x = partial pressure of oxygen (mm Hg)

Hill explained that his object was "to see whether an equation *of this type*" could account for the dissociation curves obtained in different laboratories where hemoglobin had been in various solutions such as saline or Ringer's. K and n are just parameters of the equation and do not have any physical meaning. He found close agreement between his equation and published dissociation curves with hemoglobin in different chemical solutions. Values of n ranged from just under 2 to just over 3, and the parameter has become known as the Hill coefficient or Hill constant. As long as n is greater than one, the equation above will give an S-shaped curve and the position of the inflection point (where the curve becomes steeper) is easily found. It describes a *positively cooperative reaction*, where a macromolecule (such as hemoglobin) shows an increasing affinity for binding a small molecule or ligand (such as oxygen) after the initial link is made. If $n = 1$, the ligand-binding reaction is said to be "non-cooperative" and proceeds at its own pace, regardless of how many ligand molecules are taken up.

Hill's proposal that hemoglobin molecules might aggregate or stick together was a misconception borrowed from the colloid chemists, but it did allow him to make a seminal contribution. Fifteen years later, another Cambridge physiologist, Gilbert Adair, showed that the hemoglobin molecule consists of four chains bound together with a molecular weight of 66,700 (Barcroft's dialyzing technique resulted in the separation of the chains). In 1970, Max Perutz, who spent a long career in Cambridge largely devoted to unraveling the atomic structure of hemoglobin in order to understand its function, proposed a detailed stereochemical mechanism that explains the Bohr effect.[46] He described how the oxygenation of hemoglobin is accompanied by subtle, dynamic, structural changes between the individual chains of the molecule. Perutz begins his discussion of the positively cooperative reaction with a biblical reference that captures the positively cooperative reaction between oxygen and hemoglobin, *"To him who hath shall be given,"* before quoting Hill's equation.

Although it does not accurately describe the underlying physiological events, Katz also defended Hill's original equation as "a useful theoretical 'half-truth,' cutting corners and oversimplifying the real situation, but still enabling one to gain some insight and to make practical, if only approximate, calculations":[47] just the sort of mathematical solution that a Wrangler would be expected to provide. Indeed, its simplicity and flexibility have proved durable attractions for biological scientists from many disciplines. The equation has been developed and applied across a remarkably wide range of problems. The Hill equation, or a variant thereof, is commonly used to analyze quantitative drug-receptor interactions and potential drug toxicities, and to model nonlinear drug-response relationships in pharmacology; in physiology, to build gating models of ion channels (the pores in cell membranes that allow selective movement of sodium, potassium, and calcium ions, for example); and in molecular biology, to examine the regulation of gene transcription.[48]

In October 1910, Trinity College awarded Hill a coveted, five-year research fellowship at in recognition of his outstanding work. The news came by telegram while he was visiting the Blundell's School Mission in Rotherhithe: "HAIL FELLOW OF TRINITY, LANGLEY." The Mission was a community program serving London's docklands; Hill oversaw the setting up of a shooting range for local youth. He had obviously decided to repay some of the generous institutional support that he received during the course of his own education. Trinity College ran a similar mission in Camberwell that was enthusiastically supported by the Master, Montagu Butler, who saw it as "bringing the young men of Trinity face to face, heart to heart, with the poorer classes of London."[49] Hill gave a series of lectures there the previous winter on physiology, inheritance, and nutrition: admission was free, questions were invited, but children were not admitted.[50]

As soon as he had read the telegram from Langley, Hill jumped onto his bicycle and rode back to Cambridge for the college admission dinner. His friend Charles Darwin wrote from Manchester, where he had just taken up his first academic post as an applied mathematician in Sir Ernest Rutherford's physics department. Darwin predicted AV would become a prize bore at Trinity High Table, never missing a free meal.[51] A more sober note came from the philosopher Bertrand Russell, saying he was "very glad indeed of your success" and pointing out that: "It is remarkable that all the people elected were originally mathematicians. I have great belief in mathematics as a groundwork for other subjects and yesterday's result confirms my belief."[52]

Hill first met Russell through Alfred North Whitehead, the mathematics don at Trinity. Whitehead arranged a five-year lectureship for Russell, his coauthor of *Principia Mathematica*, to teach the principles of mathematical logic. Hill and Russell as fellows found themselves on the same staircase in Nevile's Court, in quarters flanked by Wren's Library at one end and the Elizabethan College Hall at the other. Russell confided to a lady friend that despite the beauty of the surroundings, his rooms were "rather severe."[53] Russell did have a toilet adjacent to his set of rooms and invited Hill to make use of it to avoid a cold stone staircase up to the one he was allotted. Although he found Russell "very kind and amusing," Hill never really trusted him. Russell told him once that his [Russell's] "convictions were in fact quite simple, but he never put them forward because that would not give him any scope for argument and wit."[54] Most of Russell's friends, including Maynard Keynes, were associated with the Apostles—a group of brilliant but flippant men who did not share Hill's conventional values.

In October 1910, Langley wrote an encouraging note to Hill, praising him for his latest excellent research work and recommending that he should go to Germany after Christmas "both to learn German and to see other laboratories."[55] German physiology laboratories were better equipped than their English counterparts, and there was a tradition, dating from the days of Helmholtz in the mid-nineteenth century, of cross-fertilization of ideas from the physical sciences. He arrived in Germany in the New Year. As in England, he traveled mostly by bicycle, starting in Leipzig and Jena, where he witnessed some fencing duels between students. The typical German student did not impress the tall, athletic, English visitor. He described them to Fletcher as "too fat, ugly, smug and covered with gashes."[56] He rode nearly 500 km southwest to Tübingen, where he spent two months in the physiology department of Karl Bürker, who shared his interests both in muscle and hemoglobin. Bürker studied the heat produced during muscle contraction. His success in measuring tiny quantities of heat owed much to his colleague Friedrich Paschen, the professor of physics in Tübingen. Paschen had recently described a series of infrared line spectra emitted from hydrogen gas. In order to study the red end of the spectrum, Paschen was expert in assembling thermopiles—series of different metals soldered together that produce tiny electric currents when subject to temperature differences. His other vital instrument was the Paschen galvanometer to detect those small electric currents. It consisted of a mirror suspended on a quartz fiber about 1 mm in length and tiny magnets, weighing about 5 g, all enclosed in an iron tube to shield it from the earth's magnetic field. Paschen gave Hill a sensitive galvanometer he had constructed to take back to Cambridge. There is no record of Hill visiting Heidelberg, nor of a meeting on that trip with Otto Meyerhof. On his return to Cambridge, Hill would immediately incorporate the Paschen galvanometer into his experimental apparatus for measuring the tiny amounts of heat generated by muscle activity.

Hill received two letters from Charles Darwin while he was in Germany. In the first, Darwin was anxious about his essay for the Smith's Prize (the top mathematical prize open to recent Cambridge graduates). Darwin's father, Sir George, who had been instrumental in Charles landing the job with Rutherford in Manchester, tried to recuse himself as an examiner but was forced to step in because J.J. Thomson was unable to understand the Lagrange calculus that Darwin junior employed in his paper. Charles revealed his anxieties as a novice university teacher to Hill, saying that he found lecturing a strain and worried that he bored his audience. On the brighter side, he wrote: "*I have come to the conclusion that the fact you don't understand a thing yourself in no way prevents you from instructing in it. I did not understand thermodynamics at all to start with, but somehow I think I do now.*"[57] Darwin's second letter, addressed to Mein Lieber Hügel, confessed that the whole of his output for that term could be "included well on 5 sheets of paper." He then went on to say:

> I have also been having a great time doing rather amusing and quite easy mathematics for Rutherford. I think he really is now finding out about what goes on in the atom. He (and I in my humble way with him) is very much pleased because he has got a theory directly contradictory to J.J.; and if it turns out right it will be rather a score.[58]

It did turn out right: Rutherford's physical intuition that atoms contain a central nucleus superseded J.J. Thomson's whimsical "plum pudding" model.

At the end of 1913, Hill applied for a new academic post, the H.O. Jones Memorial Readership in Physical Chemistry. The origins of this post were truly tragic: Humphrey Owen Jones FRS, a fellow of Clare College, and his young bride were killed on their honeymoon in 1912 while climbing in the Alps. Hill's father-in-law, John Neville Keynes, was the Registrary or senior administrator of the university and so was ideally placed to advise him on how to optimize his application. His colleagues in the physiology school lent powerful support too. Hopkins wrote a letter stating that Hill enjoyed such a high reputation for research that any further appraisal of his talent was "as unnecessary, as it would be impertinent" and that he was "a natural teacher, most charming of colleagues with great and satisfactory influence over students."[59] Barcroft described his "mastery of calorimetry, mathematical treatment of nerve and muscle excitation and his analysis of reactions of certain colloids," burnishing his claim to be a chemist by adding that Hill had recently given a series of lectures on the application of the Second Law [of Thermodynamics] to colloid reactions.[60] Several others brought up his outstanding mathematics ability and suggested this would be an invaluable tool for the future of physical chemistry.

Hill crafted an application letter to the General Board of Studies that cleverly reinforced all of the above. He stated that his interests from his early mathematics training had always been in physics and physical chemistry, "and Biology chiefly appeals to me as a subject affording Physical and Physico-Chemical problems of the greatest interest and importance."[61] This was, in fact, an accurate description of the motivation for his future scientific career. The Readership was his.

Notes

1. A.V. Hill, Notes on Nobel Prize AVHL II 5/36/19.
2. E.H. Starling to Muriel Patterson (10/12/22) AVHL II 4/78.
3. E.H. Starling to Muriel Patterson (10/12/22) AVHL II 4/78.
4. A.V. Hill, "Autobiographical Sketch," *Perspectives in Biology and Medicine* 14 (1970): 27–42; revised by A.V. Hill in *Memories and Reflections* (1974), 673–712 AVHL I 5/6
5. School reports and letters AVHL II 5/105.
6. archives.balliol.ox.ac.uk/History/blundells.asp (accessed 22/10/19).
7. School reports and letters AVHL II 5/105.
8. Hill (1974).
9. Hill (1974).
10. Letters from W.W. Rouse Ball to Ada Hill 14/3/05 and 1/7/05 AVHL II 5/25.
11. W.M. Fletcher to A.V. Hill (11/9/05) AVHL II 4/57.
12. A.V. Hill, "History of the Hills" AVHL II 5/9.
13. A.V. Hill, Pocket diary 1905–1906 AVHL II 5/8.
14. W.M. Fletcher to A.L. Francis (2/10/05) AVHL II 4/57.
15. A.V. Hill, "On Things One Didn't Ought" AVHL II 5/25.
16. A.V. Hill, "An Adventurous Life" M&R 126–129. www.chu.cam.ac.uk/archives/collections/memoirs-v-hill/.
17. Andrew Warwick, *Masters of Theory: Cambridge and the Rise of Mathematical Physics* (Chicago: University of Chicago Press, 2003), 273–274.
18. Warwick (2003), 282.
19. School reports and letters AVHL II 5/105.
20. G.P. Thomson, "Charles Galton Darwin, 1887–1962," *Biographical Memoirs of Fellows of the Royal Society* 9 (1963): 69–85.
21. Diary AVHL II 5/23, CAC.
22. W.W. Rouse Ball to A.V. Hill (11/6/07) AVHL II 5/105, CAC.
23. G.H. Hardy, *A Mathematician's Apology* (Cambridge: Cambridge University Press, 1967), 79.
24. R.A. Herman to A.V. Hill (28/7/07), AVHL II 5/24.
25. W.M. Fletcher to A.V. Hill (2/8/07) AVHL II 4/57.
26. A.M. Hill to A.V. Hill (undated 1907 letter) AVHL II 5/28.
27. Malcolm Longair, *Maxwell's Enduring Legacy* (Cambridge: Cambridge University Press, 2016), 56.
28. J.N. Langley, "Sir Michael Foster—In Memoriam," *Journal of Physiology* 35 (1907): 233–246.
29. Robert E. Kohler, "Walter Fletcher, F.G. Hopkins, and the Dunn Institute of Biochemistry: A Case Study in the Patronage of Science," *Isis* 69(3) (1978): 331–355.
30. W.M. Fletcher, "John Newport Langley—In Memoriam," *Journal of Physiology* 61 (1926): 1–15.
31. H.M. Butler to A.V. Hill (6/5/08) AVHL II 5/24.
32. H.M. Butler to A.V. Hill (23/6/08) AVHL II 5/24.
33. A.V. Hill, "The Third Bayliss-Starling Memorial Lecture: Bayliss and Starling and the Happy Fellowship of Physiology," *Journal of Physiology* 204 (1969): 1–13,
34. Fletcher (1926).

35. J.N. Langley, "On the Reaction of Cells of Nerve-Endings to Certain Poisons, Chiefly as Regards to the Reaction of Striated Muscle to Nicotine and to Curari," *Journal of Physiology* 33 (1905): 374–413.

36. A. Maehle, "Receptive Substances: John Newport Langley (1852–1925) and His Path to a Receptor Theory of Drug Action," *Medical History* 48 (2004): 153–174.

37. A.V. Hill, "Reminiscences of the Physiological Laboratory at Cambridge in 1909-10," *Journal of Physiology* 212 (1971): 1–6.

38. A.V. Hill, "The Mode of Action of Nicotine and Curari Determined by the Form of the Contraction Curve and the Method of Temperature Coefficients," *Journal of Physiology* 39 (1909): 361–373.

39. B. Katz, "Archibald Vivian Hill, 26 September 1886–3 June 1977," *Biographical Memoirs of the Fellows of the Royal Society* 24 (1978): 71–149.

40. J. Gunawardena, "Biology Is More Theoretical than Physics," *Molecular Biology of the Cell* 24 (2013): 1627–1629.

41. D. Colquhoun, "The Quantitative Analysis of Drug-Receptor Interactions: A Short History," *Trends in Pharmacological Sciences* 27 (2006): 149–157.

42. H.P. Rang, "The Receptor Concept: Pharmacology's Big Idea," *British Journal of Pharmacology* 147 (2006): S9–S16.

43. J. Barcroft and A.V. Hill, "The Nature of Oxyhaemoglobin, with a Note on Its Molecular Weight," *Journal of Physiology* 39 (1909): 411–428.

44. J.B. West, "Three Classical Papers in Respiratory Physiology by Christian Bohr," available at journals.physiology.org/doi/pdf/10.1152/ajplung.00527.2018.

45. A.V. Hill, "The Possible Effects of the Aggregation of the Molecules of Haemoglobin on Its Dissociation Curves," *Journal of Physiology* 40 (1910): iv–vii.

46. M.F. Perutz, "Stereochemistry of Cooperative Effects in Haemoglobin," *Nature* 228 (1970): 726–738.

47. Katz (1978).

48. S. Goulette, M. Maurin, F. Rougier, et al., "The Hill Equation: A Review of Its Capabilities in Pharmacological Modelling," *Fundamental & Clinical Pharmacology* 22 (2008): 633–648.

49. trinitycollegechapel.com/camberwell (accessed 12/2/20). Trinity in Camberwell, London, is still functioning as a community center more than a century later.

50. AVHL II 5/117.

51. C.G. Darwin to A.V. Hill (11/10/10) AVHL II 5/5.

52. B. Russell to A.V. Hill (11/10/10) AVHL II 5/5.

53. R.W. Clark, *The Life of Bertrand Russell* (Harmondsworth: Penguin Books, 1978), 160.

54. A.V. Hill "Bertrand Russell (1872–1970), An Adventurous Life" in M&R 126–129.

55. J.N. Langley to A.V. Hill (8/10/10) AVHL II 4/51.

56. A.V. Hill to W.M. Fletcher (undated) AVHL II 4/27.

57. C.G. Darwin to A.V. Hill (7/2/11) AVHL II 5/24.

58. C.G. Darwin to A.V. Hill (4/3/11) AVHL II 5/24.

59. F.G. Hopkins (31/12/13) AVHL II 4/45.

60. J.S. Haldane wrote to Hill on 19/11/13 saying, "Barcroft told me that you'd been getting some new and very pretty results on the thermodynamics side of the association and dissociation of carbon monoxide and oxyhaemoglobin. Haemoglobin is rather wonderful stuff!" AHVL I 3/29.

61. A.V. Hill (8/1/14) AVHL II 4/79.

3

Otto Fritz Meyerhof

By 1820, the Meyerhof family had been living south of Hanover in the small city of Hildesheim for at least one hundred years and was well established in the textile trade.[1] Home was in the small Jewish quarter of Lappenberg. The Jewish community comprised only about 1 percent of the town's population, which was predominantly Lutheran. The Meyerhofs had long enjoyed the same civic rights as the majority population. In 1840, Otto's grandfather, Israel, began manufacturing fustian—a hardwearing, cotton-based fabric. Israel was entrepreneurial and also worked as an insurance broker for Italian and Hanoverian companies. He married Therese Gumpel, the daughter of a textile manufacturer in nearby Brunswick. In 1854, he decided to move to the larger city of Hanover, about 30 km away, with Therese and their five children, after selling his fustian factory to his brother-in-law.

Israel bought a three-story stone house in the center of Hanover, not far from the Leine river that meanders through the city. Initially he entered a business deal with a company run by the May family in Hamburg, but after a few years he opened his own manufacturing and wholesale textile business, specializing in plush fabrics such as silk and velvet. Therese would bear him five more children in Hanover, in addition to playing an active charity role supporting poor Jewish families. Israel was also prominent in public affairs, even beyond the Jewish community, as a member of the chamber of commerce and sitting on the commercial court. He befriended Hermann Wilhelm Bödeker, a Lutheran pastor, who was the city's greatest generator and distributor of charitable funds. The two oldest Meyerhof sons, Felix and Wilhelm, became partners in their father's firm. In 1881, Felix married nineteen-year-old Bettina May. Bettina was the well-educated daughter of David May, the successful businessman from Hamburg with interests in textiles and insurance. The young couple set up home on the triangular Theaterplatz in a second-floor apartment. It was a lively area with popular cafés and trams rumbling below their canopy-shaded windows. Their first child, Therese, was born there in 1882. Otto, their first son, followed on 12 April 1884. Needing more space, the family moved to a house on Georgstrasse— still in the heart of the city—where a third child, Walter, was born in in 1886.

Until 1866, Hanover was a separate kingdom, but during the Austro-Prussian war, it was annexed by Prussia. Although not popular with the citizenry, the change was dynamic for trade: the population of the city tripled in the two decades after the Israel Meyerhof moved there. He died in 1885, and his second son Wilhelm (who married Bettina's sister, Toni) moved to Berlin, where he believed the commercial opportunities would be even greater. Felix and his young family soon followed, and the brothers

Bound by Muscle. Andrew Brown, Oxford University Press. © Oxford University Press 2023.
DOI: 10.1093/oso/9780197582633.003.0003

established "I. Meyerhof" as a specialist textile company. Berlin had been the capital city of the new German Empire for about fifteen years, and, driven by industrialization and its elevation in status, was undergoing an unprecedented expansion. Much of the old city was demolished so that new streets could be laid, and new apartment buildings and offices built. The public health of the city was immeasurably improved by the construction of a new sewer system. Mark Twain, visiting Berlin in 1892, caught the essence of the modernization: "It is a new city; the newest I have ever seen. The main mass of the city looks as if it had been built last week."[2] Twain was impressed by the wide, straight streets—brilliantly lit at night—and the orderliness of life there.

In this modern metropolis, with a population of around 1.5 million, the sons of Israel Meyerhof turned the company founded in his name into one of the leading textile concerns in a fabric district dominated by Jewish businesses. The Meyerhofs lived in the central Tiergarten area, which was one of the most desirable neighborhoods with large villas and generous gardens. Their last child, Paul, was born there in 1894. Tiergarten was popular not only with businessmen but contained many embassies, as well as the homes of leading professors of the arts and sciences. When Otto was old enough for secondary school, he attended the prestigious Royal Wilhelms Gymnasium. At the time there were about 1,000 students (including many Jewish boys) in the school that was close to the Potsdamer Platz. Otto Kübler, the director, was a classicist, and the school produced many noted scientists and mathematicians as well as writers, musicians, and politicians.

The Wilhelms Gymnasium was in the tradition of humanistic, academically rigorous schools that prepared its boys for subsequent university education. Otto studied the classics, mathematics, and modern languages, as well as art and philosophy. Other technical high schools were being opened in order to produce students more in tune with the needs of an industrial society. Otto's formal schooling was interrupted at the age of sixteen years, when he developed an inflammatory kidney disease,[3] acute glomerulonephritis, probably as the consequence of a streptococcal throat infection or scarlet fever. The hallmark of glomerulonephritis is the leaking of excessive amounts of protein into the urine, causing puffy swelling of the face and limbs, as well as increased blood pressure. Otto experienced weakness and other symptoms of anemia, needing many months on bed rest. His mother, Bettina, was used to having domestic staff, and while they performed some of the nursing duties, she decided that her most important role would be to keep her clever son mentally stimulated. Bettina had a great love of the arts and encouraged Otto to develop similar interests. He read voraciously, especially the works of Goethe, and began to write poetry of his own. Bettina's nurturing love was rewarded with Otto's intellectual blossoming as well as an exceptionally close relationship with her son.

As Otto gradually recovered his strength and renal function, his physician recommended the great panacea of sunshine. His cousin, Max Meyerhof, had come to Berlin to be with his sister whose husband had died suddenly. Max qualified in Strasburg as a physician in 1897 and made a promising start to his research career by publishing a paper on diphtheria.[4] In Berlin, he was unable to find a position as a

Figure 3.1 Otto Meyerhof (left) and Max Meyerhof (right) in Egypt, 1900
Photograph courtesy of Marianne Emerson and the Meyerhof family.

bacteriologist and instead volunteered as an assistant in an eye clinic. As a physician he would be the perfect companion for the recuperating Otto, and the cousins got on well despite the decade difference in age. The pair were well matched—both were interested in history, anthropology, and archeology. They set off by train and boat for Egypt in November 1900 (Figure 3.1): the five-month trip was a defining experience for Max, who was appalled by the prevalence of blindness in Egypt due to trachoma.[5] Inspired by Max's example, Otto decided to study medicine after completing his time at the Wilhelms Gymnasium in 1903.

Although his secondary education had been interrupted by illness, the habit of broad study, cultivated by his mother, meant that Otto was in fact well equipped for life as a medical student. The best assessment of German medical schools in the early twentieth century is a report produced by Abraham Flexner for the Carnegie Foundation of New York in 1912.[6] Flexner recognized: "The German theory holds that the disciplinary period ends with the *Gymnasium*; that for a profession, once chosen, responsibility must rest squarely on the individual."[7] He could not help noticing the mobility of German medical students. Not tied to one institution, they could migrate freely in search of clinical experience that might be unavailable in their school of origin. If a university hospital in a smaller city lacked, for example, a children's clinic, the student could spend a semester at a larger institution in Berlin or Munich: "Every German student can thus piece together for himself a clinical experience in which, so far as quantity of material or quality of teaching is concerned, there is absolutely no defect whatsoever."[8] Meyerhof followed this pattern, studying in Freiburg, Berlin, Strasburg, and Heidelberg.

For the most part, Flexner was impressed by the teaching in German medical schools, where he found a uniformity of approach based on up-to-date science. The all-male, academic staff specialized in a variety of biological sciences. In a large center such as Berlin, anatomy lectures and demonstrations were given in amphitheaters overflowing with hundreds of students. There would be two professors of anatomy and only six to eight assistants. Dissections were carried out in four dissecting rooms, with fifty students in each supervised by a single instructor. Elective courses such as comparative vertebrate anatomy, embryology, and anthropology were offered, and these gave the keenest students the chance to shine. Physiology as a subject had separated from anatomy in the 1850s; departments were mostly purpose-built, with a lecture hall and associated prep rooms, as well as laboratories set up for research. As with anatomy, learning was based on high-quality lectures and practical classes. Flexner regretted that students could matriculate just through passive attendance, but in order to impress they had to demonstrate self-reliance and competence. Flexner believed that a more or less bungled experiment in a student's own hands was more valuable than the most brilliant demonstration from a professor. There was the opportunity for the energetic student to engage in research projects, and even extracurricular activities. While in Berlin, Meyerhof also established night classes for laborers that continued for many years.[9]

Meyerhof became friends with a boy named Leonard Nelson while growing up in Berlin. Nelson, two years older than Meyerhof, was the son of a wealthy Jewish lawyer. He was a driven, controversial, and contradictory character, who was completing a PhD in philosophy in Berlin in 1904.[10] The subject of his dissertation was the work of Jakob Fries—an obscure, neo-Kantian philosopher from the early nineteenth century. Fries (and Nelson) believed that Kant's "critique of reason" could serve as the foundation for a philosophy of natural science. Nelson founded a Neo-Friesian School to disseminate and teach his ideas; Meyerhof became the editor of its journal, *Abhandlungen der Friesschen Schule*.[11] This extracurricular activity brought Meyerhof into contact with many rising intellectuals and philosophers, "like-minded and tremendously disciplined young people,"[12] and with Nelson's patron at the University of Göttingen, the great mathematician, David Hilbert.

Meyerhof completed his clinical courses in Heidelberg, a city of only 50,000 inhabitants but with a tradition of medical teaching going back 500 years. One notable, progressive step was the admission of women medical students in 1900, although the predominantly male culture still centered on beer drinking and dueling. The teaching institutes that cultivated and taught the medical sciences in Germany were closely attached to hospitals. Each hospital had a central pathology department and clinical specialties housed in their own pavilions. There was one professor of internal medicine and one professor of surgery at Heidelberg, each with about one dozen assistants. The core subject that captured Meyerhof's interest as a student was psychiatry. The head of the department was Franz Nissl, who was also a talented microscopist and neuroanatomist. Patients with psychiatric or neurological disorders would be admitted to the designated pavilion in the hospital grounds, which served as a much more enlightened setting for research and teaching than the old insane asylums. Turnover was high, with about 2,000 patients occupying 100 beds during the course of one year.[13] A contemporary of Meyerhof's who carried out research in psychiatry after qualifying in 1909, Karl Jaspers, recalled the collegial spirit in the department: "It was a remarkable world of mutual spontaneity with an awareness we all shared of participating in a tremendous expansion of knowledge, with all the arrogance of those who knew too much, but also with a kind of radical criticism that subverted every position."[14]

Meyerhof received his doctorate in medicine from Heidelberg in 1909. His thesis was on the psychological aspects of mental illness; as a resourceful editor, he also published it in the philosophical journal of the neo-Friesian school. In the same volume, he included his fifty-page essay on Goethe's studies of natural science—a subject that would continue to fascinate him throughout his life.

His philosophical activities brought him into contact with a petite, dark-haired young woman, Hedwig Schallenberg, who had come from her home city of Cologne to Heidelberg to study mathematics and physics—rare subjects for a female student in 1908. Her father owned a grocer's shop in Cologne. Hedwig, the youngest of five children, was the only one to pursue further education and took courses at various German universities. There is a photograph of her as the only woman in a gathering of neo-Friesian acolytes in Darmstadt in 1911 (Figure 3.2). She is standing at the opposite end of the group from the handsome, mustachioed Otto, who seems to have eyes only for his friend Leonard Nelson, but things would change over time.

A new professor of medicine, Ludolf Krehl, was appointed at Heidelberg in 1907. Flexner described German academics as essentially wanderers, as they had been since the Middle Ages: "A student here, an assistant there, a docent elsewhere, finally a professor somewhere else."[15] The meritocracy of the German system, where "performance alone obtains results" and "inbreeding is utterly unknown," impressed Flexner. Krehl certainly conformed to that type. After beginning his career in Leipzig, he worked for short periods in Jena, Marburg, Tübingen, and Strasburg before arriving as director of the medical clinic in Heidelberg, where he would spend the rest of his career. National recognition flowed from his textbook, *Pathologische Physiologie*, that would eventually run to fourteen editions. In addition to interests in cardiology and endocrinology, Krehl was fascinated by the psychological aspects of diseases. Meyerhof's lucid doctoral dissertation must have made him an attractive junior recruit to Krehl's team. Aside from clinical duties in the pavilion, Meyerhof was expected to take part in laboratory work, and there he met Otto Warburg.

Warburg was from a distinguished and talented family of academics and bankers. He had already obtained a PhD in chemistry under the supervision of the most

Figure 3.2 Nelson's philosophy school, Darmstadt 1911. Leonard Nelson (2nd from left), Otto Meyerhof (4th from left), David Hilbert (left of center with grey beard). Hedwig Schallenberg is the only woman present
Photograph courtesy of Marianne Emerson and the Meyerhof family.

famous organic chemist in Europe, Emil Fischer. Fischer won the Nobel Prize for Chemistry in 1902 and made many fundamental discoveries in protein chemistry and enzymology. He described the peptide bond that links amino acids together and also demonstrated the specificity of enzymes in cleaving or hydrolyzing sugar molecules. Warburg, for his PhD in Berlin, investigated the enzymatic hydrolysis of proteins and how it depended on the precise, three-dimensional shape of molecules.[16] He came to Heidelberg to complete a medical degree and continued his biochemical research. He was an assured and exceptionally skillful experimenter, who now focused on the chemical energy needed to sustain biological growth. His model was the egg of the sea urchin, chosen he said because "much happens in a short time." Warburg's theory was that chemical work is necessary as biological material is added through growth after fertilization of the egg. In a series of papers beginning in 1908, he showed that oxygen consumption increases up to sixfold upon fertilization and was completely inhibited by minute amounts of cyanide and also by narcotic drugs.

Warburg took Meyerhof under his charge and began to teach him the techniques to study respiration, oxygen consumption, and growth rates in sea urchin eggs. The pair formed a productive team, with Meyerhof soon providing ideas of his own. Warburg had already made several visits to the Stazione Zoologica di Napoli (SZN), an international research center for marine biologists.[17] Sea urchin eggs, freely available from purple urchins in the Bay of Naples during the spring breeding season, had been the favorite model for embryologists to study since Richard Hertwig in 1875 had for the first time been able to observe the process of fertilization, as a spermatozoon penetrated into the transparent egg and their two nuclei fused as one. Meyerhof joined Warburg for several trips to Naples and delighted in meeting other researchers from around Europe, and in the creative exchanges that followed. Meyerhof's first research paper as an experimental scientist originated from work at SZN. He measured the heat production from sea urchin spermatozoa,[18] calculating that one billion sperm would yield almost 0.5 calories per hour. Meyerhof completed this research during the first few months of 1911. Meyerhof was seeking a research position back in Germany and wrote to Warburg, who was in Berlin, to say he had turned down a position in a municipal hospital, where there was little opportunity for "pure scientific work"[19] and he would be expected to cover for other junior physicians on the wards.

For the first few months of 1912, Meyerhof was in Naples on his own, with Warburg again in Berlin. Meyerhof wrote him long letters about his experiments.[20] Warburg was intent on uncovering the chemical and physical basis of respiration in cells. By respiration, he meant the series of chemical reactions that was enabled by catalysts to cause the oxidation of nutrient substances at biological temperatures, ultimately generating water, carbon dioxide, and energy. In 1912, he and Meyerhof published a paper on the respiration of killed cells and cell fragments.[21] For the experiment, they ground the eggs to a pulp using sand and found that respiration continued for a limited time afterwards. In the case of fertilized eggs, grinding abolished the amplification of respiration normally seen, and they behaved like fragments from unfertilized eggs. Most of the work was done by Meyerhof in Naples, where he had been expecting Warburg to rejoin him, but Warburg never left Berlin. Warburg seems to have been struggling with his latent homosexuality, and a sympathetic Meyerhof recommended physical and psychotherapy.[22] The joint authorship of their paper, mediated by frequent letters, Meyerhof described as tortuous. He complained to Warburg:

> Although I could have restructured the work, to make it more effective, if I had previously discussed things with you in greater detail—which sadly has been prevented by physical distance—I find it rather unsatisfactory for you to take such a completely negative attitude, and one which is in part dialectic.[23]

Warburg's theories of cell respiration were challenged by Heinrich Wieland, an organic chemist in Munich, who believed that oxidation in biological systems could occur without oxygen through the chemical removal of hydrogen atoms. Warburg took personal exception to this "theory of dehydrogenation," to the extent that he and Wieland no longer spoke to one another.[24] When Meyerhof expressed some initial interest in Wieland's theory, Warburg took it as a direct affront. Meyerhof mailed a copy of their eventual paper in November with an accompanying letter to Hill in Cambridge, addressing him as "Sehr geehrter Herr Kollege"[25] (Dear Honored Colleague). He followed this, some months later, with a visit in person to Cambridge, where he met Hill for the first time as well as other members of the physiology school.

As we shall see, this visit would have given him exposure to the research going on there into the chemical and physical aspects of muscle physiology.

In the spring of 1912, Otto Meyerhof was appointed as a *Privatdozent*, an unpaid teacher, at the University of Kiel. Bordering the Baltic Sea, Kiel had been designated as the premier base for the German navy, and the city had expanded tremendously in the previous quarter of a century. There were tens of thousands employed in Imperial and commercial shipyards that were booming as a result of Admiral von Tirpitz's policy to match the might of the Royal Navy and to project Germany's global power. The new professor of physiology at Kiel, appointed one year before Meyerhof, was Albrecht Bethe. He was a man of liberal disposition, who was among the first to recognize and demonstrate the plasticity of the nervous system. Much of his work in this regard had been carried out on crabs at the zoological station in Naples.[26] This must have been where he first encountered Meyerhof: academic positions were not advertised and generally depended on personal contacts. Meyerhof made an immediate and lasting impression in Kiel with his dissertation and inaugural lecture "On the Energetics of the Cell Process."[27] He saw himself working in the same tradition as Louis Pasteur and Jacques Loeb, exploring the chemical dynamics of the living

Figure 3.3 The Meyerhof-Schallenberg Wedding, Berlin 1914

(left to right) Karl Schallenberg (brother), Otto Meyerhof, Hedwig Schallenberg, Else Schallenberg (sister), Bettina May Meyerhof (mother), Susanna Schallenberg (sister), Therese Meyerhof (sister), Felix Meyerhof (father), David Schallenberg (father), Walter Meyerhof (brother), Paul Meyerhof (brother)

cell. The big issue for the modern biologist or physiologist was to understand by what physical and chemical means life is created and maintained. Meyerhof wanted to illuminate the series of chemical reactions in the cell that converted the energy from sugars, fats, and proteins, taken in as foodstuffs, into functions that maintained the cell in a dynamic equilibrium. After working with avian blood cells, yeasts, sea urchin eggs, and bacteria, Meyerhof decided his chosen tissue for study would be muscle because it offered the chance to correlate chemical transformations with the production of external work and heat.[28]

Although Otto's career to date had followed the typically peripatetic model for a young German medical scientist, he had not lost contact with Hedwig Schallenberg, the mathematics student seven years his junior, he first encountered in 1908. The Meyerhof family was prosperous thanks to the textile business built up by the two generations before Otto, so he did not feel financially insecure even though his university appointment carried no salary. With the prospect of a more settled life, he felt able to propose to Hedwig, and they married in June 1914 (Figure 3.3).[29]

Notes

1. W. Selke and C. Heppner, "Die Familie du Nobelpreisträgers Otto Meyerhof in Hannover," *Hannoversche Geschichtsblätter* 71 (2007): 156–66 www.researchgate.net/publication/331 984354r (accessed 6/20/20).
2. Mark Twain, "The Chicago of Europe," *Chicago Daily Tribune*, 3 April 1892.
3. R.A. Peters, "Otto Meyerhof (1884–1951)," *Obituary Notices of Fellows of the Royal Society* 9 (1954): 174–178.
4. J. Schacht, "Max Meyerhof (1874–1945)," *Osiris* 9 (1950): 7–32.
5. www.encyclopedia.com/religion/encyclopedias-almanacs-transcripts-and-maps/meyer hof-max (accessed 1/11/19) and Schacht (1950). Max kept a diary of their visit that is in the German Archeological Institute in Cairo. He resolved to specialize as an ophthalmologist and returned to Cairo in 1903, spending most of his medical career there. He also mastered Turkish, literary Arabic, and Persian, becoming a noted historian of Islamic medicine and science. One of his many papers on the history of ophthalmology concerned trachoma's devastating effect of Napoleon's invading army. See A. Mooreville, *Oculists in the Orient: A History of Trachoma, Zionism and Global Health, 1882–1973*, UCLA PhD thesis (2015) (accessed 22/8/20).
6. A. Flexner, *Medical Education in Europe: A Report to the Carnegie Foundation for the Advancement of Teaching* (Boston: Merrymount Press, 1912).
7. Flexner (1912), 76.
8. Flexner (1912), 160.
9. Peters (1954).
10. Leonard Nelson (1882–1927), www.friesian.com/nelson.htm (accessed 6/11/19).
11. D. Nachmanson, S. Ochoa, and F.A. Lipmann, "Otto Meyerhof (1884–1951): A Biographical Memoir," *Science* 115 (4/4/1952): 365–368.
12. F. Leal, "Who Was Leonard Nelson," in *L. Nelson: A Theory of Philosophical Fallacies* (Heidelberg: Springer, 2016), 6–9.

13. Flexner (1912).

14. E. Shorter, *A Historical Dictionary of Psychiatry* (New York: Oxford University Press, 2005), 125.

15. Flexner (1912), 145.

16. H.A. Krebs, "Otto Heinrich Warburg, 1883–1970," *Biographical Memoirs of Fellows of the Royal Society* 18 (1972): 628–699.

17. B. Fantini, "The Stazione Zoologica, Anton Dohrn and the History of Embryology," *International Journal of Developmental Biology* 44 (2000): 523–535. The SZN was founded by Dohrn, a wealthy German-trained zoologist, in 1872. Although situated in southern Italy, its scientific pedigree was distinctly German. The theory that fundamental unit of all living organisms is the cell was largely due to German physiologists such as Theodore Schwann (1810–1882), while the leading figure Johannes Müller (1801–58) promoted the study of marine life as key in understanding fundamental biological concepts—an approach extended by one of Dohrn's mentors, Ernst Haeckel (1834–1919).

18. O. Meyerhof, "Untersuchungen über die Warmetonung der vitalen Oxydationsvorgange in Eiern I–III," *Biochemische Zeitschrift* 35 (1911): 246.

19. O. Meyerhof to O. Warburg (28/3/11), in Petra Werner, *Otto Warburg's Beitrag zur Atmungstheorie: Das Problem der Sauerstoffaktivierung* (Marburg: Basilisken-Presse, 1996), 279–280.

20. Werner (1996), 266–279.

21. O. Warburg and O.F. Meyerhof, "Über Atmung in abgetötegen Zellen und in Zellfragmenten," *Pflügers Archiv für gesamte Physiologie* 148 (1912): 295.

22. S. Apple, *Ravenous: Otto Warburg, the Nazis, and the Search for the Cancer-Diet Connection* (New York: Liveright Publishing, 2021), 18

23. O.F. Meyerhof to O. Warburg (22/10/12), in Werner (1996), 306–308.

24. Werner (1996), 78–79. In 1926, Hill visited Warburg's laboratory with Meyerhof, when on their way to an International Physiology Conference in Stockholm. At the time Warburg was investigating the reversible inhibition of cell respiration by carbon monoxide (CO). Hill remembered that thirty years before, J.S. Haldane had shown the combination of CO with hemoglobin was reversible by light. This gave Warburg the key to discovering the respiratory enzyme now known as *cytochrome a_3* that is sensitive to CO. Warburg also stated that this enzyme was located in a specific site within the cell that he called "grana" (many years later identified as mitochondria). See A.M. Otto, "Otto Warburg Effect(s)" (2016), in www.cancerandmetabolism.biomedcentral.com/articles/10.1186/s40170-016-0145-9 (accessed 7/3/2020).

25. A.V. Hill, *The Ethical Dilemma of Science* (New York: Rockefeller Institute Press, 1960), 192–194.

26. K. Wiese, *Frontiers in Crustacean Neurobiology* (Basel: Birkhauser, 1990), 7–8.

27. O. Meyerhof, *Zur Energetik der Zellvorgaenge (Vortrag)* (Gottingen: Vandenhoeck & Ruprecht, 1913). In addition to publishing the lecture as a monograph, Meyerhof enlarged the material into an introductory text for biology students, *Chemical Dynamics of Life Phenomena* (Philadelphia: J.P. Lippincott, 1924), that was still in use in universities five decades later.

28. Nachmanson et al. (1952).

29. W. Meyerhof, *In the Shadow of Love* (Santa Barbara, CA: Fithian Press, 2002), 15.

4
Heat and Lactic Acid

The presence of heat as a sign of life has a long history, going back at least to the ancient Greeks, and probably beyond. The great medical scientist and philosopher, Galen, writing in the second century A.D., stated that the body's *innate heat*, conserved by the heart, was the wellspring of digestion, nutrition, and of the various humors or temperaments. Galen also observed that *excited motion* increases body heat. The concept of innate heat, generated in the heart (and cooled by the surrounding lungs), held sway for about 1,500 years before it was transformed, through a succession of revolutions in chemistry and physics, rather than in biology itself.[1]

In the late seventeenth century, the *phlogiston doctrine* became the widely accepted explanation for how fire is maintained, but after a century this notion was overthrown by the oxidation theory that emerged from the work of scientists such as Priestley and Lavoisier.[2] It was Lavoisier who likened respiration to slow combustion, with oxygen being consumed in the production of carbonic acid (carbon dioxide) and water. By the mid-nineteenth century, it was established that respiration, or the consumption of oxygen, is not restricted to the lungs but also takes place in muscle tissue. Paradoxically, it was also demonstrated that muscle contraction can still take place in the absence of oxygen. This led a German physiologist, Ludimar Hermann, in 1875 to propose that in muscle respiration, the uptake of oxygen and the production of carbonic acid are not directly linked. He suggested that muscle stores a complex, giant-sized, oxygen-containing molecule he termed *inogen*. It was the explosive breakdown of inogen with release of its oxygen, rather than the uptake of fresh oxygen, that Hermann suggested as the energy source for muscle contraction.[3]

Thermodynamics was the most important development in nineteenth-century physics to catalyze progress in physiology. It concerns heat and temperature, and how forms of energy may transform from one to another and do work. Although the subject was initiated by the desire to design more efficient steam engines, the first law of thermodynamics (*the law of the conservation of energy*) was formulated in 1847 by a young surgeon in the Prussian army, Hermann von Helmholtz. Helmholtz, the Galen of Germany, was a physicist as well as a physiologist and made fundamental contributions to visual and auditory perception. His 1847 pamphlet,[4] which announced the impossibility of perpetual motion machines, was also an attack on the long-held concept of *vitalism* in biology. He had carried out experiments to measure the heat generated by electrical stimulation of frog muscle and concluded that it was commensurate with muscle contraction and the energy exchanged—no additional, mysterious, vital force was needed.

Bound by Muscle. Andrew Brown, Oxford University Press. © Oxford University Press 2023.
DOI: 10.1093/oso/9780197582633.003.0004

In 1863, while professor of physiology in Heidelberg, Helmholtz gave several lectures in which he compared the idea of machine work with the manual power of man.

> Now, the external work of man is of the most varied kind as regards the force or ease, the form and rapidity, of the motions used on it, and the kind of work produced. But both the arm of the blacksmith who delivers his powerful blows with the heavy hammer, and that of the violinist who produces the most delicate variations in sound, and the hand of the lace-maker who works with threads so fine that they are on the verge of the invisible, all these acquire the force which moves them in the same manner and by the same organs, namely, the muscles of the arms.[5]

Helmholtz differentiated between the stored energy or force used by arm muscles in doing work and the heat lost to their surroundings. He described how animals take in oxygen and nutrients and, following chemical reactions, "generate in their place heat and mechanical force."

Walter Fletcher, Hill's tutor at Trinity, had been involved with experiments in muscle physiology since he was a medical student in the 1890s. At that time, the prevailing theory of chemical energetics in muscle was still that the inogen molecule broke down into simpler chemicals with the release of its oxygen. In the sixth edition of his *Textbook of Physiology*, which appeared in the mid-1890s, Sir Michael Foster, the founder of the Cambridge school of physiology, enshrined the German viewpoint for his readers (who must have included A.V. Hill during Part II of the Tripos).

> The oxygen taken in by the muscle, whatever be its exact condition immediately upon its entrance to the muscular substance, in the phase which has been called "intramolecular," sooner or later enters into a combination, or, perhaps we should rather say, enters into a series of combinations. ... We cannot as yet trace out the steps taken by the oxygen from the moment it slips from the blood into the muscular substance to the moment when it issues united with carbon as carbonic acid [carbon dioxide solution in water]. The whole mystery of life lies hidden in the story of that progress, and for the present we must be content with simply knowing the beginning and the end.[6]

The German physiologists knew from experiment that muscle contraction and muscle trauma were accompanied by the production of lactic acid and carbon dioxide, even in anaerobic conditions (where there was no detectable oxygen supply). In his early experiments, Fletcher observed that repeated contractions and relaxations of frog muscle in an atmosphere of nitrogen produced very little carbon dioxide. If the specimen was then allowed to recover in oxygen, carbon dioxide was rapidly released from the muscle. This experiment proved to Fletcher that "the contemporary and immediate supply of oxygen did affect the products due to contraction, and the inogen theory postulating a previous inclusion of oxygen within the muscle elements, was evidently inadequate."[7]

In 1907, Fletcher joined forces with Frederick Hopkins, the physiological chemist who kept the cages of smelly rats in the cellar of the physiology lab. Hopkins brought a new level of precision to the chemical analysis, both by processing the muscle specimens in ice-cold alcohol to exclude artifactually high levels of lactic acid, which had bedeviled previous experiments over the previous forty years, and by devising a new simple method for detecting lactic acid.[8] This enabled them to make repeated, accurate estimates of lactic acid concentrations that coincided with the known functional state of the prepared muscle. They found very low levels of lactic acid in resting frog muscle, whereas chemical, mechanical, or heat injury to the muscle caused a large surge in lactic acid release. If muscle was fatigued by repeated contractions due to an applied electric stimulus, they found a marked increase in lactic acid content (but still to less than half the levels found in injured muscle). If the muscle was then exposed to an atmosphere of pure oxygen, the lactic acid disappeared, rapidly at first and then slowly over a period of hours.

We have seen how Hill's first research papers (on receptor molecules in muscle cells and on the binding of oxygen to hemoglobin) had a lasting impact on science in the twentieth century that exceeded the youthful expectations and even the mature cognizance of the author himself. After this remarkable beginning, Hill concentrated on heat production in biological tissue, mainly muscle. The spark that ignited Hill's career came in a note from the head of the physiology laboratory, J.N. Langley, in November 1909 (Figure 4.1).

Langley continued:

> There is an especial problem suggested by Fletcher and Hopkins' work as to the efficiency of the muscle working with and without oxygen. . . . Once started there are plenty of further experiments to do and the question is a very important one for muscle physiology. It would be an advantage that Fletcher and Hopkins have done a good deal of work closely connected with this, so that you would have people interested in the subject to talk it over with.[9]

Langley cannot have foreseen the oak tree of scientific publications, still growing half a century later, nor indeed the worldwide recognition that would come to his young research student, as a result of the acorn he dropped into Hill's mailbox. Hill's first paper on heat production in muscle appeared in the summer of 1910.[10] Using apparatus that Langley found in a laboratory cupboard, Hill was able to record the instantaneous release of heat resulting from both a momentary twitch and more prolonged contractions of isolated frog muscle. He regarded the heat production as a direct reflection of the metabolic changes taking place in the muscle tissue.

A more substantial paper followed early in 1911, appearing while Hill was making his tour of German laboratories. With assistance from both Keith Lucas and the Cambridge Scientific Instrument Company, he made improvements to his first

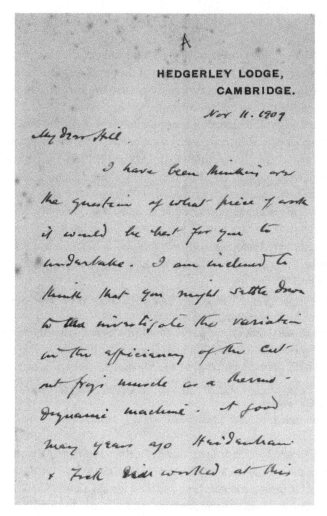

Figure 4.1 Letter from Professor Langley to his new research student, Hill, suggesting that he "settle down to investigate the variation in the efficiency of the cut-out frog's muscle as a thermodynamic machine"

Photograph courtesy of Professor Nick Humphrey/Churchill Archive Centre, Cambridge (Papers of A.V. Hill, AVHL).

apparatus in order to be able to resolve the time course of his observations in more detail. The thin specimen of muscle from the frog leg was kept at constant length so that stimulation of the muscle resulted in "isometric" contraction: increased tension in the muscle fibers not shortening of the fibers. With this novel experimental setup, Hill was able to show that stimulated muscle will continue to produce heat, during the recovery phase after the mechanical process of contraction is complete. Hill's overall conclusion was emphatic: "The muscular machine is concerned with the transformations of chemical energy into the potential energy of increased tension."[11]

He also addressed the question of the thermodynamic efficiency of the muscular machine, suggested by Langley. The efficiency of a steam engine is defined as the ratio of the work it produces to the heat it absorbs. Hill recast the concept for calculating the true efficiency of muscle contraction. He proposed it to be the ratio of the potential energy "thrown into an active muscle by excitation" divided by "the total chemical energy liberated as heat." His experiments showed that the efficiency was reduced by fatigue and cold, among other factors.

After his sojourn in Germany, Hill was able to read a long review paper on the thermal phenomena in active muscle published by Otto Frank in 1904. One of the final questions raised by Frank made a deep impression: "Do the thermal phenomena in active muscle give an answer to the problem of whether two separate chemical processes are involved: one in contraction, the other, in relaxation?"[12]

The summer of 1911 was exceedingly hot and presented practical problems for Hill working in the basement of the physiology laboratory. To exclude the influence of external conditions on the tiny temperature changes that he wanted to measure, he invented a differential micro-calorimeter. This comprised two identical thermos or vacuum flasks, which were linked by a thermocouple connected to a sensitive galvanometer. One flask would contain the experimental material, the other was filled with water. He was able to show with this new instrument that a 20 g, live mouse liberates about 70 times the amount of heat per hour compared to a 20 g live frog. He also recorded the heat released as milk sours, as yeast reacts with sugar, and as saliva breaks down starch.[13] Although Hill made various efforts to ensure that the two vacuum flasks were at the same ambient temperature by shielding them from drafts and sunlight, the intense summer heat made him uncertain about the reliability of measuring the tiny temperature increments that he expected of about 0.01° C. Therefore, he delayed completion of the experiments until the following winter.

As he explained in the final paper,[14] Hill was building on Fletcher's earlier work on the release of carbon dioxide from muscle and the Fletcher and Hopkins paper on lactic acid. He summarized his investigations as an "attempt to follow the disintegration processes of a muscle cut off from its circulation from the point of view of the total change of energy in these processes." For the first few hours, the isolated frog muscle continued to show chemical activity, but at a declining rate. Hill reasoned that sufficient oxygen remained in the muscle to permit this continuation of the normal oxidative processes of life. If the muscle was exposed to extra oxygen, more heat was generated, and the production of lactic acid retarded. As the oxygen in the dying muscle was exhausted, Hill observed a second phase corresponding to the production of lactic acid in the absence of oxygen. This was accompanied by a low level of heat production due to the breakdown of a lactic acid precursor in the muscle and a steady release of carbon dioxide as the acid was neutralized. He was pleased that the curves he obtained of heat production by the slowly dying frog muscle, under various conditions of oxygenation and pH, could be approximately superimposed on the previous curves for carbon dioxide release and lactic acid production.

The "triple similarity" of carbon dioxide, lactic acid, and heat-liberation convinced him that the three processes represented three aspects of the same reaction. The essence of that reaction was the breakdown of some unknown precursor of lactic acid into an unidentified chemical product plus lactic acid and heat. Heat production, although subject to many sources of error, could be measured simultaneously with changes in the state of the muscle (unlike the time-consuming chemical assay of lactic acid production). It seemed to reflect, in real time, the underlying biochemical reactions.

Although Langley, as an editor, decried overlong, confused, wearisome papers that absorbed "scanty library funds" and stole away "the poor hours of leisure"[15] of the active scientist, he seemed to have no qualms about publishing papers of fifty pages or more in the *Journal of Physiology*. Hill attached an appendix to his already lengthy account of the heat production of amphibian muscle in order to educate his fellow physiologists on the application of thermodynamics to the chemical engine that is muscle. He reminded them of the term "free energy" (coined by Helmholtz) that, for any spontaneous chemical reaction, denotes the maximum amount of mechanical work that can be accomplished. In particular, he was betting on the chemical transformation of the unidentified precursor of lactic acid as providing the free energy for muscle contraction. He calculated that the precursor molecule could not be glucose because it would not bring enough free energy. He favored, mistakenly as it would turn out, a single chemical reaction on the grounds that the contraction and relaxation of an insect's wing muscle was of such short duration that there was not time for a series of reactions. One can imagine that this appendix, by itself, carried significant weight in Hill's successful application for the new readership in physical chemistry the following year.

After his initial experiments, Langley encouraged Hill to apply for a research grant from the Royal Society and took "a friendly interest" in his results. He edited Hill's papers submitted to the *Journal of Physiology*, and often made Hill rewrite them again, sometimes three times.[16] Hill's ready access to Fletcher and Hopkins was invaluable, and he was able immediately to review his observations with the two authors of the classic paper on the biochemistry of lactic acid.

Otto Meyerhof visited Cambridge during this period and would have discussed aspects of muscle physiology with many of the men in the crowded department. He took particular interest in Hill's myothermic work and in his use of thermodynamics as a way to understand the various processes. With Hopkins, Meyerhof had the opportunity to discuss chemical processes that take place within living cells to maintain a dynamic equilibrium necessary for life. Hopkins soon expounded his ideas in public at a meeting of the British Association for the Advancement of Science (BAAS) in September 1913. Hopkins suspected that organic chemists believed that biochemistry was bound to be inaccurate because of the large, complex molecules found in living tissue, whereas he was intent on identifying "simple substances undergoing comprehensible reactions."[17] These reactions were controlled by specific enzymes operating in polyphasic media within the cell. Traditional organic chemists, he suggested, were more accustomed to merely separating and identifying the large structural molecules comprising animal tissues. Not enough attention had been paid to the smaller, soluble molecules present in the cell juices that take part in chemical reactions catalyzed by specific enzymes. Meyerhof's inaugural lecture at Kiel in 1912 on the energetics of cell processes, with its emphasis on the application of thermodynamics to biochemical changes and its ambition to link the underlying chemistry and energy transformations to cell function, was very much in sympathy with the Cambridge school.

In 1912, Hill regarded the breakdown of the lactic acid precursor molecule as the immediate source of energy that fueled muscle contraction. By the next year, he was attributing an additional role, viz., "lactic acid in large quantities undoubtedly causes an increase of tension in the muscle fibre": lactic acid was part of the actual mechanism of muscle fibers shortening.[18] He improved the thermocouple arrangement so that he could rapidly measure rises in muscle temperature as tiny as one-millionth of a degree. He concentrated on measuring heat production from 10 seconds to several minutes after the muscle contraction was finished. During this recovery phase, in the presence of ample oxygen, there was as much heat generated as there had been during the contraction. In an atmosphere of pure nitrogen, no such heat liberation during recovery was observed. Hill believed that the heat produced during recovery reflected the reconstitution of the lactic acid precursor molecule—a process that required oxygen.

At a meeting of the Physiology Society in February 1914, Hill presented a thermodynamic argument about the fate of lactic acid. He argued that the production of 1g of lactic acid is accompanied by the liberation of about 450 calories of heat. He measured about the same heat change during the recovery phase for muscle, whereas if 1g of lactic acid is completely oxidized or burnt it releases about 3,700 calories of heat. Therefore, he concluded, in the presence of oxygen at least 80 percent of lactic acid is resynthesized to its precursor state, while perhaps 20 percent was oxidized to provide the energy for that resynthesis.[19]

Viktor von Weizsäcker was a physician and physiologist working under Professor Krehl in Heidelberg. He obtained his MD one year after Meyerhof and was also an adherent of the neo-Friesian philosophy group organized by Leonard Nelson. So, it was probably Meyerhof who suggested to Weizsäcker that he should spend some time working in Cambridge. Apart from the obvious benefits to be gleaned by discussing research with such an active group of physiologists, Meyerhof would have described Hill's technical innovations for the original Paschen galvanometer as well as his improved thermocouples for detecting tiny quantities of heat. Weizsäcker spent the early part of 1914 working in the basement with Hill's equipment but kept his muscle specimens in salt solution rather than in air so that they were viable for much longer. Hill had suggested in 1911 that while oxygen was essential for the chemical reaction and therefore heat production during recovery, it might not be needed for the initial phase of heat production that accompanied muscle contraction. Weizsäcker set out to examine that initial phase in more detail. He was able to demonstrate that when oxygen is replaced by nitrogen or oxidation is prevented with cyanide, the generation of mechanical energy is unaffected.[20] So oxygen is not required in the process of converting chemical energy into mechanical energy in muscle, but is necessary for the removal of lactic acid during the recovery phase.

The variations in muscle function and biochemistry depending on whether there was oxygen present or not strongly reminded Meyerhof of analogous results that Louis Pasteur had observed in the mid-nineteenth century with a more primitive life-form: yeast. Pasteur noted that when yeast cells are exposed to oxygen, their sugar consumption and production of alcohol slows precipitously compared with the anaerobic process of fermentation. He attributed this to a biological property of the yeast microorganism, but at the end of the century Eduard Buchner showed that the juice extracted from yeast cells squeezed in a hydraulic press was still capable of fermenting sugar. So, Buchner concluded, the fermentation process does not require "such a complicated apparatus as represented by the yeast cell"[21] and instead depends on a protein dissolved in the juice (an enzyme he designated as *zymase*). Fermentation is, therefore, a biochemical rather than a process dependent on microorganisms, as Pasteur had believed, and required enzymatic activity, not a "vital force." Further detail was provided in 1906 by two British biochemists working at the Lister Institute in London. Harden and Young showed that a soluble phosphate added to a mix of sugar and yeast juice greatly increased the rate of fermentation, albeit for a limited period of time.[22] Harden and Young also separated or dialyzed cell-free yeast juice into two fractions that are both needed for fermentation of glucose, but are inactive by themselves. The residue left after dialysis containing the enzyme, zymase, could be inactivated by heating; the dialyzable, water-soluble component was not affected by boiling. They called it a "coferment," although a more modern term is *coenzyme* (defined as a substance not capable itself of causing a biochemical reaction to take place, but an essential accomplice).

It had been apparent for some years that the old physiology laboratory in Cambridge was grossly inadequate for serving the needs of the increasing number of pre-clinical medical students, not to mention the researchers condemned to work in the basement. In 1910 the Drapers' Company of London offered £23,000 to build a new school of physiology, which was formally opened 9 June 1914. That evening there was a celebratory dinner at Trinity College, attended by Margaret Hill's parents and husband. She stayed at home, while AV spent the night in college. He arrived home the next morning to find that their first child, Mary Eglantyne, had arrived in the early hours.

Hill acquired another visiting researcher, Jakub Parnas, from Strasburg University. His short visit was not as joyful or successful as Weizsäcker's. Parnas was obese, weighing about 350 pounds, and he displayed a rigid chauvinism that Hill disliked. He had just published a paper with a colleague in Strasburg that reported that as lactic acid is formed in fatigued muscle, an equal amount of glycogen (a storage form of glucose in muscle) disappears. Now he wanted to explore the chemical fate of lactic acid during recovery. In contrast to Hill's recent conclusion that only about 20 percent was oxidized, with the remainder being reformulated into its precursor substance, Parnas determined that all the lactic acid was burnt and not rebuilt.

Meyerhof was fascinated by these glimpses into the chemical processes underlying the transformations of energy in living organisms, starting with the digestion of food-stuff and ending with the liberation of heat and carbon dioxide. How did biological oxidation, the process of combustion or respiration, inside the cell work? In chemical terms, oxidation can either result from the addition of oxygen or the removal of hydrogen. As we have seen, Heinrich Wieland, an organic chemist in Munich, alienated Warburg by suggesting that oxidation in biological systems could occur without oxygen through the catalytic removal of hydrogen atoms from substances inside cells. Wieland cited as support for his proposed mechanism the fact that many compounds derived from foodstuffs such as citrates and lactates in the presence of biological enzymes will transfer hydrogen atoms to the dye methylene blue, reducing it to a colorless liquid. He argued that methylene blue is taking the place of oxygen, and his dehydrogenation theory, published in 1912, attracted much support. Otto Warburg remained unconvinced and thought that biological oxidation depended on oxygen being activated by iron-containing particles associated with membrane structures he called "grana." He called the unknown catalyst he believed to be located on the surface of iron particles the "respiratory enzyme." Meyerhof came to share Warburg's doubts about the theory of dehydrogenation, largely based on energy considerations, even though he was more open minded on the possibility.[23]

When he came to Kiel in 1912, Meyerhof chose an apartment that was only two minutes' walk from the physiology laboratory. The department was in modern, purpose-built premises in the gardens of an old palace.[24] He and Hedwig continued to live there after their wedding in June 1914. In Kiel that summer, there were tangible harbingers of the coming war. The latest trio of huge battleships rolled down from the shipyards less than two years after their keels had been laid. The district of Düsternbrook, where Meyerhof lived and worked, runs along the western edge of the Kiel Fjord, a long inlet of the Baltic Sea. Kaiser Wilhelm II came to race his yacht at the annual regatta in the last week of June. The festivities also included the completion of the widening of the Kiel Canal so that the battleships could steam directly from the Baltic to the North Sea, avoiding the need to circumnavigate the extended thumb of the Jutland Peninsula to the north. The regatta was interrupted on June 28 by news of the assassination of the Archduke Franz Ferdinand of Austria, and the Kaiser returned to Berlin. That summer as the British Foreign Secretary, Sir Edward Grey, remarked: "The lamps are going out all over Europe." For Hill, as we shall see, there would be no more biochemistry or physiology for five years.

Meyerhof's severe case of glomerulonephritis as a teenager had left him with chronic kidney disease, so he was spared from active military service (Figure 4.2). He was called up to serve as a medical officer on the Western Front but discharged because on health grounds in 1915.[25]

Figure 4.2 Otto Meyerhof in the uniform of an army medical officer, 1914
Photograph courtesy of Marianne Emerson and the Meyerhof family.

That year his beloved mother, Bettina, died in Berlin. The following summer their first son, Gottfried, was born—a moment of rare happiness to set against the privations of war. The Royal Navy imposed an effective blockade of shipping, and there were food riots in Kiel and other German cities. A daughter, Bettina, arrived in 1918. Like many intellectuals, Otto and Hedwig decided to abandon the Jewish faith and baptize their children as Lutherans in order to ease their assimilation into mainstream German culture.[26]

Unlike Hill, Meyerhof was able to continue his research during the war and studied a variety of biological materials in his laboratory. He decided to return to the unresolved dispute between Warburg and Wieland by testing whether methylene blue accelerated the normal oxidative processes in cells. He observed its presence led to increased consumption of oxygen and production of carbon dioxide. He concluded:

> After the elimination of a certain part of the respiratory enzyme—as a result of destruction, damage or poisoning—in determined conditions methylene blue can, by means of its oxygen-carrying properties, substitute for this part of the respiratory enzyme in the oxidative mechanism.[27]

Meyerhof's experiments were not still enough to convince Warburg that oxidation via the respiratory enzyme located in grana could take place at the same time as dehydrogenation in the cytoplasm. Although he mused about this dual process in public on several occasions in public, Warburg's combative personality kept a largely one-sided dispute with Wieland going for another decade, before he convinced himself that both did occur.[28]

The profound question forming in Meyerhof's mind was, What chemical processes are responsible for the creation and preservation of life? The similarities between yeast and muscle, for example, being able to switch from respiration to fermentation suggested a biochemical unity across all forms of life. He published a trio of papers on the biochemistry of nitrogen-fixating bacteria, while important papers on muscle started to appear in 1918. Meyerhof used an improved version of a blood-gas monitor invented in England by Haldane and Barcroft to measure the quantity of carbon dioxide released during fermentation. He reported that the coferment (the coenzyme of alcoholic fermentation in yeast) is present in the muscle and organs of animals, and in milk, but is absent from serum.[29] Following the experiments of Harden and Young, Meyerhof demonstrated that by mixing the zymase-containing extract from minced muscle with the conferment or coenzyme component of yeast juice, he could still produce lactic acid from glucose. Based on these simple but elegant experiments, he ventured to suggest the coenzyme fractions of yeast and muscle were either identical or very similar. Adding to the notion of universality of biochemical reactions, Meyerhof also managed to extract the coenzyme from germinating peas. By the time these papers appeared in *Zeitschrift für physiologische Chemie*, there very few, if any, British readers of the journal.

Notes

1. E. Mendelsohn, *Heat and Life* (Cambridge, MA: Harvard University Press, 1964).
2. M. Teich, *A Documentary History of Biochemistry 1770–1940* (Rutherford, NJ: Associated University Presses, 1992).
3. Teich (1992), 126–127.
4. H. Helmholtz, "On the Conservation of Force [1847]," in J. Tyndall and W. Francis (eds.), *Scientific Memoirs Selected from the Transactions of Foreign Academies of Science* (London: Taylor and Francis, 1853), 114–162 (available in Google Books).

5. H. Helmholtz, "On the Conservation of Force (1863)," sourcebooks.fordham.edu/mod/1862helmholtz-conservation.asp (accessed 1/12/19).

6. W.M. Fletcher and F.G. Hopkins, "The Respiratory Process in Muscle and the Nature of Muscular Motion" (Croonian Lecture, 1915), *Proceedings of the Royal Society of London*, B 89 (1917): 444–467.

7. Fletcher and Hopkins (1917).

8. W.M. Fletcher and F.G. Hopkins, "Lactic Acid in Amphibian Muscle," *Journal of Physiology* 35 (1907): 247–309.

9. A.V. Hill, "The Heat Production of Muscle and Nerve, 1848–1914," *Annual Review of Physiology* 21 (1959): 1–19.

10. A.V. Hill, "The Heat Produced in Contracture and Muscular Tone," *Journal of Physiology* 40 (1910): 389–403.

11. A.V. Hill, "The Position Occupied by the Production of Heat, in the Chain of Processes Constituting a Muscular Contraction," *Journal of Physiology* 42 (1911): 1–43.

12. Hill (1959).

13. A.V. Hill, "A New Form of Micro-calorimeter for the Estimation of Heat Production in Physiological, Bacteriological or Ferment Actions," *Journal of Physiology* 43 (1911): 261–285.

14. A.V. Hill, "The Heat Production of Surviving Amphibian Muscles during Rest, Activity and Rigor," *Journal of Physiology* 44 (1912): 466–513.

15. Hill (1959).

16. Hill (1959).

17. F.G. Hopkins, "The Dynamic Side of Biochemistry," *Annual Reports of the British Association* (1913): 652.

18. A.V. Hill, "The Energy Degraded in the Recovery Processes of Stimulated Muscle," *Journal of Physiology* 46 (1913): 28–80.

19. A.V. Hill, "The Oxidative Removal of Lactic Acid," *Journal of Physiology* 48 (Suppl) (1914): x–xi.

20. V. Weizsäcker, "Myothermic Experiments in Relation to the Various Stages of a Muscular Contraction," *Journal of Physiology* 48 (1914): 396–427.

21. Teich (1992), 43–47.

22. J.N. Prebble, *Searching for a Mechanism: A History of Cell Bioenergetics* (Oxford: Oxford University Press, 2019), 43–47.

23. P. Werner, "Learning from an Adversary? Warburg against Wieland," *Historical Studies in the Physical and Biological Sciences* 28, no. 1 (1997): 173–196 Thirty years later, it was shown that the respiratory enzyme was bound to the inner membrane of mitochondria.

24. www.physiologie.uni-kiel.de/en/history (accessed 30/12/19).

25. E. Hofmann, R. Ulrich-Hofmann, and W. Höhne, "Otto Meyerhof and the Exploration of Glycolysis—Outstanding Research in an Inhumane Era," *Vorträge und Abhandlungen zur Wissenschaftsgeschichte, Acta Historica Leopoldina* 59 (2012): 317–382.

26. W. Meyerhof (2002), 20.

27. O.F. Meyerhof, "Untersuchungen zur Atmung getöteter Zellen. I.," *Pflüger's Archiv für die gesamte Physiologie* 169 (1917): 97–121, quoted in Werner (1997).

28. Werner (1997).

29. A. Harden, *Alcoholic Fermentation* (London: Longmans, Green & Co., 1923).

5

Hill's Brigands and the Great War, 1914–1918

The tumultuous days at the beginning of August 1914 were especially hectic for AV and Margaret. Their daughter was just six weeks old and beginning to stare at people—she already seemed interested in what was going on around her. She was known as Polly—Eglantyne Jebb's pet name for Margaret. On Sunday morning, 2nd August, Margaret's brother, Maynard, telephoned to say that he had been summoned to the Treasury because they wanted to pick his brains for the country's benefit.[1] Maynard Keynes believed the war would be short-lived. His parents were on holiday in France, from where his mother wrote: "I hope your faith in the European situation will be justified." But if there was going to be war, it should be run as efficiently as possible, so Maynard asked AV if he would transport him to London because he was concerned that the train service might be disrupted. Hill needed no excuse to kick-start Buster. Maynard decided it would be improper to be seen climbing out of a sidecar in Whitehall and made AV drop him off several streets away (Figure 5.1).

The onset of the Great War is often held to have been sudden and unforeseeable, but some of Hill's contemporaries at Trinity College noticed harbingers of the coming hostilities well before, if not imagining its duration and cataclysmic scale. Charles Darwin, who had been in the Officers Training Corps at school, was about to take the "A certificate" needed to become an officer in 1911. He wrote to Hill, then in Germany, to say he was "getting rather anxious about military matters." With his usual jocularity, he admitted to being shy of mixing with the "boys," but was determined to be duly qualified "to repel your present friends coming invasion."[2] Looking back to the years before World War I, Hill suggested, "it was commonly regarded as a sign of feebleness of intellect to believe that war was conceivably possible."[3] No doubt he was reflecting on the wide influence in university circles of Norman Angell's 1910 book, *The Great Illusion*, that argued war involving industrialized nations was economically irrational, and that the all-round destruction of wealth and international trade it would entail acted as a powerful disincentive.

Hill, by contrast, identified himself as a member of "a substantial minority, less intellectual but not a bit less intelligent" of young scientists, who made one of the earliest contributions to the national preparedness. The Army reservists in Cambridge already included infantry, cavalry, and artillery. In 1910, F.J.M. "Chubby" Stratton, an astrophysicist who had been 3rd Wrangler in 1904, started a company of signals. They were provided with workshop space by Bertram Hopkinson, professor

Bound by Muscle. Andrew Brown, Oxford University Press. © Oxford University Press 2023.
DOI: 10.1093/oso/9780197582633.003.0005

Figure 5.1 AV riding Buster with his sister Muriel in the wicker sidecar, Cambridge
Photograph courtesy of Professor Nick Humphrey/Churchill Archive Centre, Cambridge (Papers of A.V. Hill, AVHL).

of engineering, and began designing and constructing radio sets, suitable for use on the battlefield. Everything had to be made from scratch—there were no shops selling components. When the signal company was in the field, a dynamic soldier on a stationary bicycle generated electrical power for the sets. Stratton was an energetic but easygoing commander, who was much more interested in the inherent problems of long-distance communication than he was in military drill.[4] In April 1912, Stratton and his officers, including Hill, visited the battlefield at Waterloo; field maneuvers in the Cambridgeshire countryside became increasingly frequent.

Great Britain declared war on Germany on 4 August 1914. That day AV's sister, Muriel, wrote to him to announce: "Tom and I are going to be married this afternoon at 2.30 pm. Please come if you can and also Margaret."[5] Tom was Tom Hele, a Cambridge physiologist whom Muriel met through working at the laboratory. With mobilization the next day, Hill was called to join the Cambridgeshire Regiment in Romford, Essex. He wrote to Margaret to say he had taken up quarters in a bicycle shed, and he had acquired one mattress, two blankets, and one feather pillow for eight shillings.[6] The following day he reported that the officers had been roused at midnight and asked to take their men to Belgium. Thinking it was "a very strange scene," Hill refused "at once" because he thought "it would be sheer murder in the present state of training of some of these boys."[7] He requested fresh vegetables and

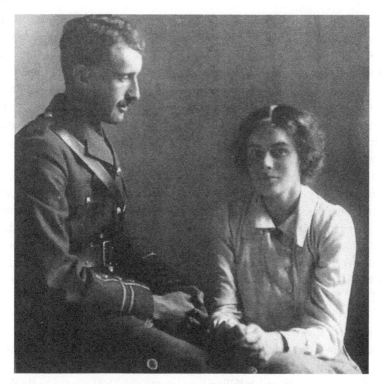

Figure 5.2 AV and Margaret Hill in 1914

Photograph courtesy of Professor Nick Humphrey/Churchill Archive Centre, Cambridge (Papers of A.V. Hill, AVHL).

fruit for his men from his "sweet lady" (Figure 5.2): Margaret sent 60 lb. of apples and plums the next day. She reported that Cambridge was filling up with soldiers, and Nevile's Court at Trinity College was being converted into a base hospital. She and her mother busied themselves finding temporary accommodation for Belgian refugees.

Hill's essential ambivalence about a war with Imperial Germany that necessarily involved friends and colleagues on opposite sides was revealed in a letter he wrote to Jakub Parnas, the Austrian biochemist who had been working in his laboratory for a few weeks. The letter was written after Britain declared war on the Austro-Hungarian empire on 12th August. Hill expressed his regrets before adding: "I hope in future Austria will cease to be an appendage of Prussia with its gospel of 'blood and iron.'"[8] He pointed out that the war had started because of Germany's "absolutely unprovoked assault on a small country" (Belgium) before saying that he had no quarrel with German-speaking people. Concerned that Parnas was trapped in Cambridge without support, he offered to lend him 25 pounds. In a postscript, he apologized for his "hard language, but remember that I have nothing but kindly feelings concerning the people we are fighting against."

Parnas showed no remorse when replying. He regretted "the acerbity of [Hill's] anathema against Germany and the Germans."[9] He did not believe press reports about German atrocities in Belgium, saying that victorious troops do not commit atrocities against conquered citizenry. Hill met Parnas several times when he came home to Cambridge, before receiving a note that he was in jail. Parnas was interned with a troupe of German circus acrobats in Warwick, before being allowed to return to Germany in November. Writing to Alexander Forbes, a young neurophysiologist at Harvard a few months later, Hill wondered if Parnas "has turned his hand to making poisonous gases"[10] for use against British troops yet.

By the end of August, Hill and his company had been transferred closer to Cambridge at Bury St Edmunds, but he was frustrated by the lack of training opportunities. "What [the men] need is MUSKETRY, of which I have given them none as yet."[11] The next day brought better news: he was going to take 100 men to Sudbury to let off 70 rounds per man. But the ammunition did not arrive until November, when he finally received "32 targets and many thousands of rounds" and his recruits "shot pretty badly." Just one week later, he told Margaret "The men are shooting very well indeed."[12]

Hill's success as a shooting instructor did not go unnoticed. He was promoted to the rank of captain and, in April 1915, made musketry officer for the 3/1st East Midlands Infantry Brigade. This meant traveling quite widely from base to base, preparing thousands of young soldiers who would soon be transferred to the Western Front. Not content with just fulfilling the training role, Hill started to think about how sniper fire might be made more lethal. He clearly believed that in close combat, any improvement that could be made to the government-issued rifle was justified. Yet in his letter to Forbes, Hill expressed revulsion about explosive bullets that the Austrian soldiers were using that caused "terrible mutilations." Even so, he admired the fiendish design of the bullet, making a drawing for Forbes and explained: "When the bullet hits, the lead weight jumps forward, hits the little striker, which detonates the explosive, and then the victim goes to bits. Nice bit of devilry, isn't it? All this on behalf of *Kultur*!"

Hill designed a prototype sight that could be attached to a standard Enfield rifle, for greater ease and accuracy when firing at distant targets. He devised a dual-sight system: a front sight on the end of the barrel with a painted ring for aiming and a rear sight that slid up and down a post. Whenever he had the chance to visit Cambridge at the weekends, he would discuss the invention with engineers at the Pye Instrument Company who manufactured the sights. In October 1915, he had opportunity to try the device for himself and, when it was almost too dark to see, he fired twenty shots into a target 100 yards away with a spread of just a few square inches. He received reports from France in December that there were concerns about the robustness of the sights under rough handling in the muddy trenches, but an officer reported "my snipers have now tested it and are delighted. ... It is far simpler and magnifies more than several of the other similar ones we have."[13]

That letter from Major Crumm was passed on by Horace Darwin FRS, another son of Charles Darwin the naturalist, and a member of the advisory panel of the new Munitions Inventions Department (MID).[14] The MID was intended to be a clearinghouse for ideas or inventions put forward by the military or indeed by the public to aid the war effort. The disaster of the Gallipoli campaign, combined with a shocking shortage of shells for the British artillery on the Western Front, had forced the formation of a coalition government in May 1915. David Lloyd George was put in charge of a new Ministry of Munitions[15] and came to view the war as "a conflict of chemists and manufacturers."[16] Horace Darwin was an engineer, who already served on a national committee for aeronautical research;[17] now he set up the Anti-Aircraft Experimental Section (AAES) at MID. In the early months of the war, pioneering airmen on both sides mostly confined themselves to reconnaissance missions over the battlefields, but it was clear that air warfare would soon become an additional offensive threat.

Horace Darwin's peacetime career was as the founder and chairman of the Cambridge Scientific Instrument Company, which supplied equipment to the physiology laboratory, among others. Darwin knew Hill through this connection and also as a friend of his nephew, Charles. Horace's only son, Erasmus, had been killed in April 1915 in the second battle of Ypres. Before the war, Horace made original contributions to the development of aeronautical instruments that would function during vibration and while accelerating in three dimensions. He now began to think about AA training and proposed a scheme that would be economical in terms of shells fired as well as measuring errors made; he wrote a report on "Aiming Practice with Anti-Aircraft Guns" in December 1915. The only AA guns in Britain were naval guns, designed for other purposes. There were three novel operational problems to be solved: (1) the position and motion of the target aircraft relative to the gun; (2) the direction in which the gun should be aimed; (3) bursting of its projectile to cause maximal damage to the target.[18]

Darwin remembered past interactions with Hill when he came to order physiology equipment in Cambridge; he thought of him as showing "signs of the unpleasant habit of inventing things."[19] At New Year 1916, Hill was recuperating from a severe bout of influenza at home in Cambridge. He received a letter written from the MID saying that his "name has been mentioned ... by Mr. Horace Darwin, and to ask if you could arrange to come up one day next week to see him."[20] Although he was not well, Hill rode Buster up to London to meet Darwin. Horace told him about his brother, Leonard, a former professional soldier, who had introduced a photographic technique in 1889 for artillery practice against a moving target such as a ship.[21] Horace thought that it might be possible to use a camera obscura to project the horizontal movements of an aircraft and combine them with its altitude gauged by a theodolite, a surveying instrument. The Central Flying School at Upavon in Wiltshire had a camera obscura that they were using to improve bombing accuracy, and it was agreed that Hill should spend a week there in February.

Upavon is on the edge of Salisbury Plain—its grass runways were converted from horse gallops. Although its primary function was to train new pilots for the Royal Flying Corps, an Experimental Flight was established to test new machines and equipment, much of which was developed at the Royal Aircraft Factory in Farnborough. Hill reported to the chief experimental officer at Upavon, an Oxford chemist named Henry Tizard.[22] Tizard, like many of his scientific colleagues at Farnborough, decided that they could offer helpful advice only if they learned to fly themselves. Tizard was allowed to fly only on days when the weather was too bad for the trainee pilots, and he had just been allowed to fly without supervision. He was involved in work to check the accuracy of a bombsight and to record the track of an aircraft flying over Upavon field using synchronized observations from two cameras obscura set up about one mile apart. One of the cameras employed a vertically mounted lens above a fixed horizontal mirror; in the other the lens and mirror were tilted.

Hill came away from his week at Upavon convinced that a surveyer's theodolite would be useless for estimating aircraft height because it could not be read or adjusted fast enough. He also decided that the type of camera obscura employed at Upavon was too bulky to move and its field of view was too restricted.[23] But he was intrigued by the idea of a horizontal mirror with an eyepiece at a fixed distance above it, as a method to plot an approaching plane. If two flat, horizontal mirrors were used that had a calibrated grid engraved on them, and they were deployed a large distance apart, observers would be able to determine the coordinates of an aircraft in three dimensions.[24] Looking through an eyepiece above the mirror, the observer would mark the reflected image of the target in the sky with a dot of glycerine ink, synchronously with the observer at the second mirror. Hill rigged up a prototype mirror system while at Upavon and derived the necessary geometrical equations[25] to generate an aircraft's 3-D trajectory from the two sets of mirror plots. He reported his ideas to Horace Darwin, who was familiar with optical recording methods, having designed a system in 1884 for Kew Observatory to measure the height and speed of clouds, using two cameras with a plotting table and strings.[26] Darwin was impressed, and the Darwin-Hill Aerial Position Finder was born. Once they were convinced that their position finder would work, Hill ordered three pairs of mirrors to be made in Hatton Garden, London, to the following specifications: glass, 22 inches square, edges rounded, silvered but to have silver scratched off by a fine tool to leave a grid of squares. When ready, the mirrors were to be shipped immediately to the Cambridge Scientific Instrument Company, where they would be mounted and have the eyepieces attached (Figure 5.3).

Hill had not been offered a permanent fellowship at Trinity when his research fellowship ended. Whether or not he was disappointed to separate from his old college, his full commitment to the war effort meant that he was out of sympathy with the antiwar movement led by some dons at Trinity. This had begun quietly with a letter to the local press in August 1914 from the non-political physicist, J.J. Thomson, who wrote that a war with Germany would be "a sin against civilization." Bertrand Russell was the most notable and active war critic, producing a stream of articles and

Figure 5.3 Hill testing the Darwin-Hill aerial position finder with Ralph Fowler
Photograph courtesy of Professor Nick Humphrey/Churchill Archive Centre, Cambridge (Papers of A.V. Hill, AVHL).

giving many antiwar speeches.[27] He also supported the Union of Democratic Control (UDC), a national pressure group that wanted an open review of the decision to go to war and opposed the introduction of conscription and consequent infringements of civil liberties.

The event that caused the greatest uproar within Trinity centered on an empty room. John Littlewood, who continued to be G.H. Hardy's closest mathematical collaborator and was also a good friend of Russell's, volunteered his services to the Royal Artillery and was commissioned as a second lieutenant. Littlewood gave his permission to Hardy to hold an antiwar meeting in his vacant rooms, a move that outraged the College Council and further split the fellowship.[28] Hill was fully aware of these developments because he wrote about them in another letter to Alexander Forbes at Harvard, in December 1915.

Cambridge is now filled with a sad lot of cranks, who believe the war is a very sad thing and ought to be stopped. ... The so-called Union of Democratic Control is run by some pure mathematicians (and others of the same kidney) at Trinity, who fancy themselves cleverer than their fellows; they held meetings in, and wrote letters addressed from, the college.[29]

Around the same time, Hill also wrote to his former tutor, Walter Fletcher, who had been appointed the first secretary of the new Medical Research Council in 1914. Telling Fletcher that he had dined at Trinity, Hill was scathing about their former colleagues: *"They do seem absurd people there now, with their feuds about Russell and other things that don't matter."*[30]

Despite these antipathies, on his return to Cambridge from Upavon, Hill beat a path directly to Trinity College, specifically to the study of the leading mathematics don, G.H. Hardy. When Hill became a junior dean at Trinity in 1912, Hardy wrote to him to say that he had voted against the appointment in a minority of one, not because he disapproved of Hill but because he objected to the post.[31] One of the main duties of the junior dean was to enforce regular attendance at chapel. Hardy and Hill then combined in a campaign to end compulsory attendance, but were roundly defeated by the Master, the Rev H.M. Butler. Hill now approached Hardy to identify some outstanding young mathematicians to carry out analytical work for the Anti-Aircraft Experimental Section (AAES). Hardy's indignant response to Hill's request was that he would not "prostitute his brains"[32] by using mathematics to enhance warfare. Hardy "deplored the fact that J.E. Littlewood had joined the army and was working with the Ordnance Committee in calculating range tables for guns." Unflustered, Hill persisted and asked if there were any younger mathematicians at Trinity College he might approach. Hardy reluctantly mentioned two: Ralph Fowler and E.A. Milne.

Ralph Howard Fowler became a research fellow in mathematics at Trinity College in 1914, after showing outstanding ability as an undergraduate. Raised in a relaxed, affluent family, supported by servants, he was very much a youth of the golden Edwardian age. He had been head of school at Winchester College and, robustly built, shared his family's sporting prowess in a variety of ball games. Yet in early 1916, Fowler was still on sick leave from a serious thoracic wound sustained at Gallipoli the previous June. As a gunnery officer in the Royal Marine Artillery, he was responsible for devising range tables for a previously untested, gargantuan, weapon: a 15-inch howitzer that fired a shell weighing 1,400 lbs. This experience was ideal preparation for the sort of analysis that AAES needed. In contrast to Fowler, Arthur Milne was a short, bespectacled, second-year undergraduate at Trinity College. His parents were both god-fearing, primary schoolteachers in Hull; as with Hill, his path to the mathematical Tripos had depended heavily on winning scholarships. When he came to sit for the Trinity scholarship examinations as a seventeen-year-old in 1913, it was his first trip outside his native Yorkshire.[33] But there was nothing modest about his mathematical talent. Having been turned down by the university OTC because of his poor eyesight, Milne volunteered as a stretcher-bearer for the wounded brought to the makeshift field hospitals in Cambridge. The prospect of applying his brain to active defense issues was irresistible.

In addition to seizing the two mathematicians most strongly rated by Hardy, Hill wrote to the "deplorable" Lt. John Littlewood, who was busy at the Ordnance Department devising an easier way to calculate the trajectories of shells fired at a high angle of elevation. Before the war, the analysis of artillery ballistics was rudimentary.

Artillery targets on the battlefield were fixed and often within sight, so that adjustments could be made on the fly. AA fire is obviously more complex, with a target moving in three dimensions. Traditional range tables applied to relatively low angles of fire, 10 to 15 degrees above horizontal. An Italian military mathematician, Francesco Siacci, had devised workable formulas for such flat fire in 1888.[34] AA fire requires elevations above 45 degrees. For such high-angle trajectories, the only method of mathematical analysis depended on laboriously integrating the path of the shell from hundreds of calculated "small arcs," where air resistance was assumed to stay constant for each arc. Air resistance remained an imprecise consideration, complicated by the decrease in atmospheric density and the strong winds at high altitude. And whereas conventional artillery just focused on producing an explosion on a fixed target, for effective AA fire the entire trajectory of the projectile needs to be known. This is necessary to set a fuse-timer that would produce an explosion close enough to damage the aircraft.

Littlewood devised a rapid and powerful method of calculating high-angle trajectories, avoiding the long process of "small arc" integration. His solution[35] was an ingenious combination of the time-height relation for "vertical fire" (the shell traveling straight up and down, encountering reduced drag in thin air) and then using this in conjunction with Siacci's formula for "flat fire" (horizontal) to determine trajectories at any angle between 0 and 90 degrees. Littlewood schooled Fowler in his new method of ballistics analysis.

Hill arranged to begin the first field trials of the Aerial Position Finder at Northolt aerodrome to the west of London. The two horizontal mirrors stations, to be set up about one mile apart, had to be linked by telephone so that the observers could synchronize their observations. He needed to find a man with experience of laying telephone wires; a colleague at the National Physical Laboratory recommended his brother-in-law, William Hartree. Hartree was in his mid-forties, and his innate modesty and shyness made it easy to underestimate his varied abilities. He had progressed from the mathematics Tripos (18th Wrangler in 1892) to engineering. He worked uncomplainingly in the engineering laboratory at Cambridge for twenty years, before deciding to retire to Surrey to amuse himself constructing wireless sets. With the outbreak of war, he volunteered to work for the General Post Office as a linesman, but now readily agreed to join Hill at Northolt, seeing it as a more direct contribution to the war effort. Hill described the living arrangements at Northolt to Margaret:[36]

> Fowler and I are together in a comfortable room. Poor little Milne has fixed himself with a policeman after a very tiring struggle to find a lodging. Hartree turned up today. He is very nice and extremely diffident. … He was a Major Scholar of Trinity so there are four past or present major Sc's of Trinity here![37]

The small group was soon ready to test its position finder. The first plane's coordinates were plotted at 3–5-second intervals on the mirrors: the readings showed that it was climbing. A second plane was instructed to fly at a constant altitude over the field, and it was measured as being at 3,840 ft., 3,820 ft., and 3,800 ft.—the pilot's instruments recorded a steady height of 4,000 ft.[38] At the end of March, three senior figures from the MID made a site visit to Northolt and were very impressed with the progress made in just a few weeks. Hill, together with Fowler, Milne, and Hartree, suggested different applications for the mirror position finders: training gun crews (without actual firing), testing gun sights, showing the position of shell bursts to check range tables, and generating high-angle range tables.[39] On Good Friday, a large crowd gathered to watch a naval gun firing blank, 3-inch smoke shells at a small plane. While the visual display was exciting, Hill realized that the calculations involved were too laborious to give meaningful feedback to the gunners.[40]

Hartree applied himself to improving the existing devices for measuring aircraft height so that they were practical for use in the field. His solution was rigged up using cables and wooden poles to give a system that gave estimates without the need for trigonometry. With Hill's help, they assembled a prototype that proved easy to use, and twenty Hartree-height-finders were sent to the army in France in May 1916.[41]

These early successes came despite some bureaucratic setbacks. Hill reminded the authorities in the spring of 1916 that he had received no pay or allowances for two months, and that Milne was no longer receiving his scholarship money since leaving Trinity. The most serious threat came when Fowler's sick leave ended. The Admiralty decided to appoint him as an inspector of steel, a mundane task for which he had no particular qualification, and his transfer to MID was repeatedly refused. By now, Hill realized that Horace Darwin was a man of extensive influence and asked him to intervene. He contacted ex–Prime Minister Arthur Balfour, another Trinity man and friend of the Darwin family, who was now First Lord of the Admiralty. Fowler's official transfer to MID was approved without further delay.[42] For Fowler, there was no sense of celebration—his younger brother, Christopher, was killed while fighting in France in early April.

The limitations of the Northolt site became insuperable after a few weeks. There was no possibility of firing live shells, and it was even difficult to arrange flights because of the "apathy from the people at the aerodrome, who regarded us a set of rather uninteresting cranks, wanting to make a science of a thing which was intended by nature to remain a sport."[43] The situation improved after an invitation to carry out experimental work at the National Physical Laboratory (NPL) in Teddington, where there was a wind tunnel. William Hartree's eldest son, Douglas,[44] had just completed his first year reading for the mathematics Tripos at St Johns College, Cambridge. He came to visit his father at the NPL and was immediately drawn into the research. Hill quickly spotted his outstanding ability and persuaded him not to return to Cambridge after the summer vacation—he became known as Hartree II. Having lost his fellowship at Trinity, in March 1916 Hill was pleased to be elected to a fellowship at King's College, an appointment no doubt aided by marrying into the Keynes family. At his first dinner at King's high table, he persuaded a senior mathematics don, Herbert Richmond FRS, to come and join the AAES, without pay. Hill was constantly looking for able recruits.

The Royal Navy had decided to defend London from Zeppelin raids by installing a 6-pounder gun on the roof of the Admiralty. The officer in charge, Commander "Barmy" Gilbert, got to hear about Hill's activities and invited him to make a visit.[45] Gilbert had invented his own "peculiar sort of anti-aircraft gun sight mounted on a 6-inch gun in a monitor [a shallow-draft gunboat] at Great Yarmouth." Gilbert was anxious to know about the accuracy of the 6-inch gun firing at a high angle of elevation, so Hill and his team arrived on the estuary of the River Yare in June. There were not many holidaymakers there in the summer of 1916, and Hartree I rigged up telephone wire along lampposts on the front at Gorleston, but they kept breaking in the gusty winds. Hartree habitually dressed like a tramp, wearing an old overcoat and battered hat, and sucked on an unlit Woodbine cigarette. One morning he was working on lines along the beach and was arrested and detained at bayonet point for an hour, which amused him. There was little productive activity during the three weeks around Yarmouth because of the English summer weather. The first day that the rain stopped and clouds lifted was a Sunday, but the captain of the monitor refused to fire his gun on the sabbath. Finally, there was a clear weekday and the Darwin-Hill Aerial Position Finder proved that for fused shells fired from the monitor gun at a fixed angle of elevation, the point of explosion bore no relation to the official range tables. They did, however, fall exactly on the trajectories predicted by Littlewood.

Commander Gilbert thought the experimental results were significant enough to send Hill to report them to senior navy officers at Portsmouth. They were so impressed that Hill was instantly invited to move his research group to HMS *Excellent*—a stone frigate of the Royal Navy Gunnery School on Whale Island in Portsmouth harbor. The commander, Vincent Lewin Bowring, was an affable sea dog, who divided his time between running a small holding there and looking after 2,000 officers and men.[46] He claimed that the roar of gunfire stunted the growth of his plants and upset the chickens. Nevertheless, he welcomed the influx of young undergraduates and Cambridge dons and named them "Hill's Brigands."

Hill wrote a happy letter to Margaret shortly after arriving on Whale Island. Two permanent mirror stations had been established, one at Eastney and the second four miles away on Hayling Island. He wrote:

> We are getting very sunburnt sitting in the sun all day by the sea and looking in the mirror. They treat us very well here. . . . I think they won't let us pay for our Messing.[47]

The frustrations of Northolt and Yarmouth were soon forgotten, and the Navy on Whale Island gave the Brigands complete cooperation. On the first clear day, Cdr. Bowring had a 3-inch gun fire smoke shells at an unusually high quadrant elevation.[48] Out of a total of nineteen shells, fourteen were "blind": they did not explode. The handful that did, exploded much later in their flight than expected. Now that they had access to plenty of ammunition, the Brigands attacked the problem with gusto. Milne wrote to his younger brother about the unresolved puzzle:

> Altogether 150 rounds were fired, but the remarkable thing was that over 50 of them were "blind," i.e. the fuses did not go off. About the remaining 100 bursts, every blessed thing is known: actual position, time of flight, time taken by sound to reach observers, deviation from angle of sight, wind blowing at the moment in the neighbourhood of the burst, elevation of gun, fuse-length, barometer, thermometer, and finally the muzzle velocity of the gun.[49]

The fuse in an artillery shell consisted of a ring of compressed gunpowder that ignited as the projectile accelerated down the barrel. When the burning surface reached a preset, critical point, it would cause the shell to explode. The rate of blind shells with conventional "flat" artillery fire was low, but the Brigands found that with the gun barrel elevated above 45 degrees, unexploded shells "fell about in an unseemly and dangerous manner."[50] Firing at high angles means that the shell loses velocity sooner than a shell fired more horizontally, but hundreds of firings showed that the rate of burning of the fuse at any given time did not depend on the velocity of the shell at that moment. Instead, the rate of burning seemed to depend on the initial (muzzle) velocity of the shell.

This seemed inexplicable until Hill put forward the hypothesis that the determining factor was the angular rather than the linear velocity of the shell. To hold an accurate course, projectiles have to be made to spin by rifling the barrel: spiral grooves force the projectile to rotate as it moves along the barrel. Whereas air resistance slows the linear velocity of a shell, the rotation about the shell's axis is relatively unaffected. Hill persuaded the navy to construct two guns with different pitches to their rifling, one with more twists than the standard and one with fewer. The gun with the with the most spiral turns in its barrel led to all the shells being blind at great heights. The gun with the least twists inside the barrel performed well, with mixed results from the intermediate, service gun. Hill did not witness the crucial trial himself because he had retired to his sick bed with another episode of flu, but he derived great satisfaction from his successful prediction. Laboratory experiments followed in the engineering department of University College London, where it was shown that the rate of fuse burning decreased the faster the fuse was rotated because of centrifugal effects on the burning slag.[51] As a result of these investigations, all British AA guns were standardized to the lesser rifling.

The atmosphere on Whale Island was generally happy and relaxed, albeit with the traditions of the senior service governing behavior in the officers' mess. The Brigands, a mix of young students (whom Hill protected from being called up), older dons, and a sprinkling of wounded soldiers, such as Fowler, were all accorded honorary officer status. Despite the comfortable circumstances, for Hill there was seldom a gap of more than a week during which there was no news of friends or acquaintances from Blundell's or Cambridge being killed or wounded in action. Every so often, the loss would be deeply felt, as in October 1916 when Hill received a letter telling him that Keith Lucas (from the physiology department) had been killed in a flying accident at Upavon. He left a widow and three young sons. AV wrote a note of condolence to Alyss Lucas, who thanked him for breaking "through for me the awful crust of dull insensitivity that has choked me for several days."[52]

The young mathematician, Milne, observed that Hill was "the inspirer of most of the researches that were conducted, whilst Fowler executed them in the field."[53] Once an idea had been put forward by Hill,

> [Fowler] tossed it back at lightning speed. ... There was a thrill about the investigation—it was as though one were physically in love with the particular problem, when Fowler was present. ... He had a power of throwing himself unrestrainedly and wholeheartedly into other people's interests which is rarely exhibited; once a thing had been expounded to him and been subjected to his searching cross-examination, he could argue on level terms about it with the leading experts on it, and usually contribute to the discussion some material consideration that they had overlooked.[54]

There was a feeling that Hill sometimes did not take the suggestions of other Brigands seriously, but as one of them remarked: "You have to admit that he has more ideas than all the rest of us put together."[55] Hill fully appreciated Fowler's qualities, writing to Walter Fletcher that he was "an extremely able person and extraordinarily hard working."[56] Hill could not resist taking a dig at G.H. Hardy, who had reluctantly recommended Fowler in the first place. Saying that Fowler was "most unlike a pure mathematician who is generally lazy," Hill hoped to find him "a job after the war to get him out of Hardy's clutches and allow him to be a useful physicist & applied mathematician."

Milne himself made several important contributions to the mathematics of ballistics. Together with his old tutor, Herbert Richmond, he was able to show that the complexities of wind buffeting a shell could be represented by the "equivalent constant wind," thus making its correction a simple matter for gunners. Hill realized that Milne and the other mathematicians were spending too much of their time on tedious, repetitive calculations and decided to approach the country's leading statistician, Karl Pearson, for help. Pearson was a true polymath, with an intimidating manner. He was a eugenicist, as were several members of the Darwin family, and ran the Galton Laboratory of Biological Science at University College London.[57] The unpredictable Pearson did agree to help, and Hill wrote to Fletcher to tell him the news:

> V generous offer of the assistance of 8 skilled computers from Karl Pearson. It may turn out v valuable. My gang of brigands here have more than they can manage at present. ... We are remaking all the Range Tables for High Angle Fire for both services, in effect: they were all based on pure thought previously and like most examples of pure thought not based on observations were quite wrong.[58]

The "8 skilled computers" were mostly young women, who crunched numbers on mechanical calculators in the Galton Laboratory. Rather than send one of the famous Cambridge dons as the representative from Whale Island, Hill wrote to Pearson introducing Milne as "a nice little modest person, very clever and rather shy."[59] Milne spent the first week of the New Year explaining the raw ballistics data to Pearson and showing him how to convert it into range tables and trajectory plots. Pearson was incensed when Hill placed the initials K.P. on the first batch of printed reports: all subsequent ones were labelled G.[alton]L.[ab] to signify his group's effort.[60] Pearson worried that errors in the data and the inherent approximations would threaten the reputation of the Galton Laboratory, and at times did not seem to realize there was a war going on. After one of Pearson's frequent complaints, Hill explained to him somewhat tetchily "the whole essence of war work is making up one's mind on insufficient data; & success is won by the trained judgement which knows what compromise to make."[61]

Hill mentioned to Pearson that "Fowler and Milne are now in Scotland going around with our so-called travelling circus, recording high angle 'battle practices' with great success."[62] The traveling circuses were intended to improve the performance of AA batteries in the field. They would set up the mirrors of the Darwin-Hill system near the gunners, and one gun would then fire a smoke shell. The other guns in the battery would fire rounds at the distant puff of smoke, and then adjust their sights. Fowler led the first traveling circus to France, where he took the opportunity to liaise with French military scientists and visited the Brocq factory, which was producing a novel electromagnetic device for predicting the track of an approaching plane.

Hill spent many weeks in France himself in 1917. The extreme level of destruction was visible everywhere he went. He wrote to Margaret at the end of May about "villages that have been rooted up and pulverized and demolished. It is the most appalling devastation and a fine example of the glories of war."[63]

Back in England, Hill's responsibilities were steadily growing. He served on various subcommittees at the MID as well as at the Ordnance Committee at Woolwich arsenal and at the War Office. In the summer of 1917, he was officially appointed Director of AAES, with Fowler as assistant director in charge at Whale Island. In addition to the work at Portsmouth, Hill oversaw the experimental research on shell trajectories going on in the wind tunnels at the NPL, the engineering and computing work at UCL, the traveling circuses plus new investigations involving location of guns and aircraft by sound at Stokes Bay near Portsmouth and at Rochford. Early in 1918 he nominated several of the Brigands for honors in respect of their war work. Hill, Fowler, and Hartree I collected OBEs, and Milne an MBE. Pearson was offered a CBE but took great pleasure in refusing it. The recognition that Hill valued most was his own election as a Fellow of the Royal Society as a result of his prewar research work. His name was proposed by England's leading physiologists, headed by Langley; it was approved in 1916, but the award suspended until the end of the war.[64]

Hill continued to make regular trips to France in 1918, liaising with French, Italian, and American defense experts. During his October visit, the Allies liberated Lille, on the French-Belgian border, and Hill at last believed the war would be over by Christmas. The armistice was signed in November, but Hill had to write long reports, including a summary of the activities of the AAES, before he could return to academic life. His overall conclusion was that while the Darwin-Hill position finder did not prove that useful in tracking aircraft, the data obtained by observing artillery fire at Whale Island and subsequent laboratory tests transformed the principles of gunnery. This was echoed in the official history of the Ministry of Munitions, which stated: "It would hardly be an exaggeration to maintain that the whole science and practice of artillery has been revolutionised by the development of AA gunnery."[65] Hill came to believe, with justification, that the traveling circuses, studying the performance of artillery batteries in the field, were pioneers of operational research. He was struck that the precision of the instruments used to measure was always greater than the consistency of the material (guns, fuses, shells, etc.) that they were designed to test. He reflected that his background as a physiologist, knowing "something about eyes, nerves and hands" and also having "long experience of rifle shooting, probably helped me to design the instruments in a way that the human senses and faculties could make the best use of."[66]

Hill's long experience of target shooting enabled him to suggest a simple physical solution to Fowler, who was making no progress on an investigation into the cant or tilt with which a shell leaves a gun. The cant results from inherent asymmetries in the explosive charge as well as in the shape of the shell and the barrel, so that the shell leaves the gun at a slight angle, resulting in a corkscrew motion or "yaw" along its trajectory. Hill proposed that the yaw could be modeled by firing a shell through a series of millboard pistol targets or jump cards, each about two square feet in area. The shape of the successive holes in the cards would reveal the wobble of the shell, allowing further analysis of its trajectory. A series of these cards was erected, and 3-inch shells were fired through them on a flat trajectory. A shell moving straight along its long axis would leave a series of circles in the cards. In practice, the shells left a series of tattered oblong holes that enabled the angle of yaw or tilt to be measured at intervals of about 60 feet. The experiment resulted in two major papers, describing the motion of a spinning shell "at velocities both greater and less than the velocity of sound"[67] that became classics in the field of ballistics.

Apart from the technical contributions of Hill's Brigands, their leader impressed high-ranking figures in the military with his judgment and his ability for the clear exposition of problems. He had a knack for producing practical recommendations after applying logic and for designing gadgets that surpassed the understanding of most senior officers. Hill could see that while the modern military depended on advances in technology, it could also benefit from the objective appraisal of scientists.

Hill's natural aptitude for leadership first became apparent in the war, when he was always collegial and never sought to dominate. He was as solicitous of the troops he was teaching to shoot in 1915, as he was of senior figures from Cambridge such as Herbert Richmond and R.A. Herman, who found themselves as Brigands. At the end of the war, he sat down with many of the Brigands individually to advise them on their futures. One notable case was E.A. Milne, who did not think he could afford to return to Cambridge to complete his degree and intended to seek paid employment to support his struggling parents. Hill thought that Milne had proved himself as a mathematician far beyond the undergraduate level. He wrote to the twenty-three-year-old:

> I am quite determined that you are going to return to Trinity, so don't think you are going away elsewhere to 'start earning' … I imagine you are very nearly a certainty for a fellowship in 1919 or 1920. Herman, Fowler and I—not to mention Richmond—can speak sufficiently strongly of your capabilities … then with 250 [pounds], plus room and dinners plus commons, approximately equal to 350 plus what you can make by teaching, you will actually be better off than by going somewhere else … you will be doing what it is your right and duty to do—to help in the promotion of knowledge.[68]

Hill's predictions were borne out and Milne, despite the lack of a degree, became the youngest fellow at Trinity College. Early in 1920, Hill heard that there was an opening for an assistant director at the Solar Physics Observatory in Cambridge and immediately recommended Milne, starting him on a career as one of the most influential astrophysicists of the century.

During the early years of the war, Hill was able to come back to Cambridge for most weekends. Margaret had decided to move in with her parents in Harvey Road. A second child, David Keynes Hill, was born in July 1915. Eglantyne wrote a sardonic letter of congratulations to her old lover:

> When will Vivian begin showing him how to kill cats and has he taught him to swim yet? I'm sure they will be great pals. He will know all about calorimeters before he knows about cats.[69]

Once the AAES work began, Hill was almost never able to come home. Although Margaret had some support from her mother and various maids, she was in charge of two infants and also started keeping rabbits, chickens, and goats for food. She wrote loving letters to AV with news about the children and saying how much she missed him. When he was in Great Yarmouth in June 1916, she reminded him: "Next Sunday is our wedding day. Three good years we shall have been married, three of the happiest years of my life. God bless you my dear, dear man."[70] The letter was signed F.W., short for "Fuzzy Wuzzy." AV's letters did not convey the same direct, loving feelings to Margaret. He was inhibited when trying to express deep emotion. He did record his growing revulsion for the war, in the form of a message to his children: "It is a beastly war in which everyone gets blown up, so don't deceive yourself when you grow up and think it is a glorious or a beautiful war." This was written before he visited the battlefields in France and witnessed the horror of mechanized warfare for himself.

In September 1916, Hill wrote to say that he was too busy at Portsmouth to come home for the foreseeable future. Margaret, who was pregnant again, decided to pay a visit to Eglantyne Jebb. She was recovering from surgery to remove a goiter from her neck and living, unhappily, with her mother in Sussex. Margaret took young Eglantyne (Polly) with her and left David with his grandparents for the week. The long periods of separation began to tell on her, especially as there was some concern about the health of her brother, Geoffrey, who was working as a field surgeon in France. At the end of November, she mused to AV that "If there were several of you all married to me, perhaps one of you would come home every weekend." Despite her increased family responsibilities, Margaret did not shirk public duties, giving advice to young mothers at a local school and helping to set up a "Co-operative Food Culture Society" to encourage production. She said she represented "the backgarden-farmer element."[71] A few days later, she reminded AV "we are both fairly young and the war cannot last until we are middle aged." With the loss of domestic staff to wartime jobs, Margaret claimed she had become "rather a dab at cooking." She also showed her Keynesian streak by deciding that keeping small numbers of chickens was uneconomic and larger flocks were necessary to obtain the maximum return in food value. She made herself unpopular by "telling people that they are not doing a patriotic act in producing quantities of eggs." She concluded that "people have not an elementary idea of the meaning of money and hate believing anything they don't want to."[72] Their fourth wedding anniversary passed in June 1917, with AV noting three of those years had been the unhappiest in the world's history, but the last one had been the happiest for him and Margaret so far.

A second daughter arrived in January 1918 and Margaret wrote to the "Dearest man of mine" telling him that she and Janet were both flourishing and that she was becoming a model baby.[73] AV replied one week later to say that he had just got back to London from Glasgow and was now on his way to Paris. He saw very little of the family that spring, but in August he did manage to take Margaret on a short break to Moreton Hampstead on the edge of Dartmoor. He made the wry observation "hence Maurice"[74]—their last child who would be born 29 May 1919.

The end of the Great War was shrouded by the great influenza pandemic, and both Polly and David were afflicted. Their grandmother, Florence Keynes, wrote that they had temperatures of 105 degrees, but unlike hundreds of soldiers and civilians in Cambridge, they both survived.[75] Margaret issued a household bulletin about the children and nurse to AV and described herself, F.W., as "lovely as ever."

The pacifist element in Cambridge was given a public voice by C.K. Ogden, a classicist and linguist, who founded the *Cambridge Magazine*. Eglantyne's younger sister, Dorothy, persuaded Ogden to start devoting half of the weekly magazine to digests of news that she had translated from European newspapers, including those from Germany and Austria. By 1918, the bulk of the stories concerned the severe food shortages that mainly resulted from the blockade of German ports maintained by Royal Navy since the start of the war. Eglantyne joined forces with her sister early in 1917 and oversaw a team of two dozen translators (being fluent in French and

German herself); she was successful in recruiting well-known public figures to what was a controversial and unpopular cause. She wrote to Margaret:

> It was a combined effort, rather than anything I did for myself, which saved me from permanent invalidity. I'm hurrying up now to try and do something for somebody before I die![76]

When the naval blockade was continued after the armistice, the two sisters organized a political pressure group, the Fight the Famine Council. Maynard Keynes, who had publicly defended Ogden and his magazine, agreed to chair an economics subcommittee. He attended the Paris Peace Conference in 1919 as the chief Treasury representative of the British delegation, where he masterminded the effort to lift the blockade and prevent mass starvation in Germany.[77]

Maynard was frustrated by the intransigence of the conference delegates and resigned from the Treasury to write *The Economic Consequences of the Peace*, while Eglantyne Jebb viewed the Paris conference as "very self-important."[78] She decided that direct action was needed to prevent the deaths of millions of children in Europe. She distributed thousands of leaflets that featured photographs of starving Austrian children, for which she was arrested in Trafalgar Square and charged under the 1914 Defence of the Realm Act. Her trial in May 1919 attracted a great deal of attention which she and Dorothy, never ones to miss an opportunity, exploited a few days later by holding a rally in the Royal Albert Hall. As a result, "Save the Children" took flight.

Notes

1. Skidelsky (1994), 289.
2. C.G. Darwin to A.V. Hill (5/5/11) AVHL II 4/19.
3. A.V. Hill, *Feebleness of Mind? Or Moral Obliquity?* M&R 505–510 AVHL I 5/6.
4. Stratton eventually served as signals officers with great courage in World War I. He spent three years on the Western Front as a senior signals officer and was awarded the DSO. In March 1918 he was responsible for communications between seventeen divisional headquarters, as the German army attempted to push the British 5th Army into the sea.
5. A.M. Hill to A.V. Hill (4/8/14) AVHL II 5/28.
6. A.V. Hill to M.N. Hill (10/8/14) in Angier I (1978):125.
7. A.V. Hill to M.N. Hill (11/8/14) in Angier I (1978):125–126.
8. A.V. Hill to J.K. Parnas (undated, August 1914) AVHL II 4/66.
9. J.K. Parnas to A.V. Hill (3/9/14) AVHL II 4/66.
10. A.V. Hill to A. Forbes (24/5/15), reproduced in the *Atlantic Monthly*, October 1916. Gas was first used by the German army at Ypres about one month before the letter was written. Geoffrey Keynes, Margaret's younger brother, was serving as a surgeon in France. He wrote to Hill (27/6/15, AVHL II 4/48) saying he firmly believed in the existence of "stinkpots"— the schoolboyish argot for canisters of chlorine. He confided to AV: "If the rumour is true, it sounds much too humane a manner of killing Boches because it is quite painless."

11. A.V. Hill to M.N. Hill (30/8/14) in Angier I (1978):128.
12. A.V. Hill to M.N. Hill (1/11/14) in Angier I (1978):128.
13. Major Crumm to A.V. Hill (30/12/15) AVHL II 5/48.
14. M. Pattison, "Scientists, Inventors and the Military in Britain, 1915–1919: The Munitions Invention Department," *Social Studies in Science* 13 (1983): 521–68.
15. Martin Gilbert, *The First World War* (New York: Henry Holt,1994), 162–163
16. Pattison (1983).
17. R.T. Glazebrook, "Obituary of Sir Horace Darwin," *Nature* 122 (13/10/1928): 580–581
18. Pattison (1983).
19. William van der Kloot, *Great Scientists Wage the Great War* (Stroud: Fonthill Media, 2014), 165.
20. MID letter to A.V. Hill (1/1/1916) AVHL I 1/1.
21. Undated notes AVHL I 1/9.
22. J.K. Bradley, *The History and Development of Aircraft Instruments, 1909-1919.*(PhD thesis, London University, 1994).
23. Undated notes AVHL I 1/9.
24. William van der Kloot, "Mirrors and Smoke: A.V. Hill, His Brigands and the Science of Anti-Aircraft Gunnery in World War I," *Notes and Records: The Royal Society* 65 (2011): 393–410.
25. "History of the Anti-Aircraft Experimental Section of MID" (undated) AVHL I 1/37.
26. Bradley (1994).
27. Clark (1975), 298–378.
28. Meg Weston Smith, *Beating the Odds: The Life and Times of E.A. Milne* (London: Imperial College Press, 2013), 20–21.
29. A.V. Hill to A. Forbes (15/12/15), repr. in "The Laboratory Reacts," *Atlantic Monthly*, October 1916.
30. A.V. Hill to W.M. Fletcher (undated) AVHL II 4/57.
31. A.V. Hill, "Let His Children Be Fatherless" M&R, 382–384 AVHL I 5/5/.
32. A.V. Hill, "On Prostituting One's Brains" M&R, 124–125 AVHL I 5/4.
33. Weston Smith (2013).
34. D. Aubin and C. Goldstein (eds), *The War of Guns and Mathematics* (Providence, RI: American Mathematical Society, 2014).
35. E.A. Milne, "Ralph Howard Fowler (1889–1944)," *Obituary Notes of Fellows of the Royal Society* 5 , no. 14 (1945): 61–78.
36. M.N. Hill to A.V. Hill (22/3/16) in Angier I (1978):138.
37. A.V. Hill to M.N. Hill (22/3/16) in Angier I (1978):138.
38. van der Kloot (2011).
39. MID report March 1916 AVHL I 1/10.
40. A.V. Hill, "History of the AAES (1918)" AVHL I 1/37.
41. van der Kloot (2011).
42. A.V. Hill, "On Wrangling in a Good Cause" M&R, 503–504 AVHL I 5/5.
43. A.V. Hill AVHL 1/37.
44. Douglas R. Hartree FRS (1897–1958), mathematical physicist and computer pioneer.
45. A.V. Hill, *The Royal Navy Club* M&R, 515–521 AVHL I 5/6.
46. van der Kloot (2011).
47. A.V. Hill to M.N. Hill (9/8/16) in Angier I (1978):141.
48. van der Kloot (2011).

49. E.A. Milne to G. Milne (18/8/16) in Weston Smith (2013), 32.
50. A.V. Hill, "How Littlewood Could Not Make Ballistics Respectable" M&R, 525–529 AVHL I 5/6.
51. van der Kloot (2011).
52. A. Lucas to A.V. Hill (24/10/16) AVHL II 4/55. Another grievous loss in a flying accident came in August 1918. Bertram Hopkinson FRS, the professor of engineering at Cambridge who played a leading role in technical developments for the RFC and RAF, crashed in bad weather over London.
53. Milne (1945).
54. Milne (1945).
55. J.G. Crowther, *Fifty Years with Science* (London: Barrie & Jenkins, 1970), 15.
56. A.V. Hill to W. Fletcher (14/12/16) AVHL II 4/57.
57. G.U. Yule and L.N.G. Filon, "Karl Pearson (1857–1936)," *Obituary Notices of Fellows of the Royal Society* 2, no. 5 (1936): 72–110. Francis Galton himself nominated Pearson to be the inaugural professor of the laboratory, in no small part because he was an outspoken eugenicist.
58. A.V. Hill to W. Fletcher (14/12/16) AVHL II 4/57.
59. A.V. Hill to K. Pearson (31/12/16) quoted in Weston Smith (2013), 37.
60. A.V. Hill, "Karl Pearson" M&R, 42–44 AVHL I 5/4.
61. A.V. Hill to K. Pearson (4/3/17) quoted in T.R.V. David, *British Scientists and Soldiers in the First World War* (PhD thesis, University of London, 2009), 140.
62. van der Kloot (2014), 184.
63. A.V. Hill to M.N. Hill (26/5/17) in Angier I (1978):149.
64. EC/1918/14 Royal Society Collections.
65. Weston Smith (2013), 34.
66. A.V. Hill, undated note AVHL I 1/9.
67. R.H. Fowler, E.G. Gallop, C.N.H. Lock, and H.W. Richmond, "The Aerodynamics of a Spinning Shell,"' *Philosophical Transactions (A)* 221 (1920): 295–387 and Part II, A222 (1922): 227–247. Richmond also compiled a two-volume "Textbook of AA Gunnery," published by HMSO (1924–1925) and used extensively during World War II.
68. A.V. Hill to E.A. Milne (25/11/18) in Weston Smith (2013), 57.
69. E. Jebb to M.N. Hill (undated) AVHL II 5/116.
70. M.N. Hill to A.V. Hill (13/6/16) in Angier I (1978):140.
71. M.N. Hill to A.V. Hill (1/12/16) in Angier I (1978):145.
72. Angier I (1978):146–147.
73. M.N. Hill to A.V. Hill (20/1/18) in Angier I (1978):152.
74. A.V. Hill, "History of the Hills" AVHL II 5/9.
75. Angier I (1978):155.
76. Mulley (2009), 216.
77. Skidelsky (1994), 358–363.
78. Mulley (2009), 231–243.

6

Postwar Tensions

Walter Fletcher and Frederick Gowland Hopkins delivered the Royal Society's Croonian Lecture in December 1915 on "The Respiratory Process in Muscle and the Nature of Muscular Motion."[1] It comprised a detailed review of advances made in the physiology and biochemistry of muscle since the beginning of the century, to which they had made such notable contributions. The opening section of their lecture included these prophetic words:

> We must still look to the study of muscular motion as the most fruitful, and perhaps for some time to come the only, avenue to intimate knowledge of the modes of energy discharge in the living cell and of their relation to the specific chemical processes of life.

They paid generous tribute to the recent work of Mr. A.V. Hill, whose investigations into heat production by muscle "has confirmed and further illuminated our own observations made on the chemical side." When the lecture was published, Hill wrote to Fletcher from HMS *Excellent*, saying that he had forgotten all the physiology he ever knew and was unable to understand an article on metabolism that J.S. Haldane had just sent him. Despite his loss of command of the subject, Hill hoped to retain the general ability that "Barcroft calls the faculty of seeing beyond the visible horizon."[2]

Hill employed the same excuse to Joseph Barcroft himself, when he attempted to draw Hill into a friendly difference of opinion he was having with Haldane about hemoglobin. Hill answered that since he had read no recent papers on hemoglobin and had forgotten the older literature, it would be impudent to dispense his opinion.[3] Barcroft was unconvinced and wrote again the next day complaining that William Bayliss, a well-known physiologist at UCL, had also questioned his contention that oxyhemoglobin is a true molecule. Barcroft could see only two alternatives to resolving the matter:

> (1) The Oxford approach and just state "Undoubtedly oxyhaemoglobin is a chemical compound" or
> (2) I can try to consolidate my present belief by seeking fresh proof that can be put before Bayliss.[4]

But Barcroft did not want to spend the remaining twenty years of his scientific life "hammering at it unless there is a prospect of settling it." After further discussion,

Bound by Muscle. Andrew Brown, Oxford University Press. © Oxford University Press 2023.
DOI: 10.1093/oso/9780197582633.003.0006

they decided to invite Bayliss to Cambridge to debate the difference of opinion. The outcome was the founding of the Haemoglobin Committee of the Medical Research Council (MRC) that, under Barcroft's chairmanship, stimulated research on hemoglobin in Great Britain over the next decade.[5]

Hill finally returned to the Cambridge physiology lab in May 1919. The war had aged him—he was now, at the age of thirty-three years, iron grey—yet, spared the worst of the conflict, he retained optimism about the future. Although he was happy with his new fellowship at King's, he still felt a past debt to Trinity for the help and opportunities it had afforded him. He wanted more contact with his old college. He complained to Walter Fletcher, now running the new MRC, about the inadequacy of the scientific staff at Trinity College, especially in the field of chemistry.[6] Hill brought back to Cambridge with him the former Trinity scholar William Hartree, who had enjoyed his time as a wartime Brigand so much that he jumped at the opportunity to work in Hill's laboratory. Hartree was fifty years old and had never published a scientific paper before. He precariously balanced his chair on four brass fuze cases, collected during the recent conflict, and was happy to work for long hours and no salary.[7] He was, of course, a strong mathematician and skillful engineer, who had no trouble mastering Hill's electronic instruments. Despite having no prior experience as a physiologist, Hartree soon became adept at dissecting frogs to obtain viable muscle and nerve preparations. Within weeks, papers from the pair started to appear on more sensitive thermocouples and better recording techniques; these were followed by a substantive analysis of the myothermal response associated with a prolonged isometric contraction. Reminding their readers that profound advances in muscle physiology depended on "incomparably greater accuracy and speed of instrumentation," they identified four phases of heat production: an initial one as tension developed in the muscle; maintenance heat during the period of maintained tension; relaxation heat at the end of stimulation; and oxidative recovery heat which continued for several minutes.[8]

One day when Hill was working in his laboratory, an unannounced visitor "blew in" and asked, "Are you Hill?" When AV confirmed that he was, the man said, "Would you like to come to Manchester?" to which Hill replied, "God forbid, why should I?" The man turned out to be H.R. Dean, professor of pathology at Manchester, who had been charged with recruiting Hill to be the Brackenbury Professor of Physiology at the university. The previous incumbent, William Stirling, had been an unproductive occupant of the chair because of a drink problem.[9] He had recently been forced to resign after insulting the wife of French president Poincaré at a reception in Paris. Hill was very uncertain about accepting the position but was swayed by Sir Ernest Rutherford, who had just moved from Manchester to become the director of the Cavendish Laboratory. Rutherford's advice was characteristically blunt: "Cambridge is a good place to be young in and it is a good place to be old in, but in the middle of your life *clear out!*"[10]

Rutherford's own successor as professor of physics in Manchester, W.L. Bragg, also weighed in after Dean told him that Hill was hesitating. Bragg, who knew Hill from Trinity College and through his own war work in France, set out to reassure him that he had many of the same reservations about moving to Manchester, but "now I am here I am just so thankful that I decided to come that I cannot express it properly. I think it is being responsible for one's own department that makes all the difference."[11] A further complication arose when Hill received a telegram from Johns Hopkins University in Baltimore, offering him a chair in physiology with an annual salary of $7,500. Hill disclosed this offer to Manchester and wrote to his old tutor, Fletcher, saying "You have always been my fairy godmother: what do you think about it?"[12] It was also a difficult decision from a family point of view since Margaret had spent her whole life in Cambridge—her parents were there, and she was currently looking after four children with whooping cough. Hill drove a hard bargain with Manchester and agreed to accept the position only after they promised "a large sum of money to re-equip the laboratory and provide a decent staff."[13]

The professor of physiology in Kiel, Albrecht Bethe, moved to the University of Frankfurt in 1915 and was succeeded by his friend and near contemporary, Rudolf Höber. Like Bethe, he was an amiable man and an accomplished scientist: his research centered on the electrical potentials recorded across cell membranes, especially in nerve and muscle cells. He was able to secure Meyerhof's promotion to assistant professor on his return from France. Just as Kiel displayed the portents of war in 1914, its ravaged economy four years later heralded the near collapse of the German homeland. There had been food riots in 1916 and unrest throughout the last two years of the war. In early November 1918, there was a full-scale mutiny by sailors who were unwilling to expose themselves to a final battle against the Royal Navy[14] that their senior officers craved. Hundreds were arrested but then released after mass protests; the red flag of revolution was raised over the warships in the harbor. The sailors immediately spread out across Germany fomenting unrest that ultimately forced the abdication of the Kaiser. The populist slogan in Kiel was "Frieden und Brot" (Peace and Bread). Meyerhof was relatively unaffected by the unsettled and depressed state of the city and spent long hours working in his laboratory. Hedwig was at home with three-year-old, Gottfried, and the baby, Bettina (Figure 6.1). To help with childcare, she hired a young woman, Anni, who had spent the war in China as governess to the family of the German military attaché.[15]

Meyerhof set out to make a definitive study of the fate of lactic acid as muscle recovered in oxygen. Hill had suggested that while some of the lactic acid was burnt, releasing energy, about three-quarters of it was reconstituted into a precursor molecule. Parnas, on the other hand, had claimed all the lactic acid was completely oxidized—a determination he had originally made in Hill's lab in Cambridge in the summer of 1914 and then published again in the German literature in 1915. Meyerhof now demonstrated beyond doubt that during the recovery period, some lactic acid was converted back to glycogen. Moreover, the energy consumed in the oxidation of lactic acid balanced exactly the deficit in resynthesized glycogen. As Meyerhof wrote: "The glycogen content at the end of the recovery period is the same as that before the beginning of the stimulation minus the carbohydrate disappearance equivalent to the

Figure 6.1 Hedwig and Otto Meyerhof, Kiel c.1920
Photograph courtesy of Marianne Emerson and the Meyerhof family.

oxygen consumption."[16] He pointed out that his energy balance measurements verified those of Hill and Hartree. This was the first demonstration of the cyclic nature of energy transformation within cells. Some weeks earlier, Meyerhof reported that under anaerobic conditions lactic acid was produced in proportion to the duration of contractile activity.[17] For the first time, a direct link was established between the work produced by muscle and the cyclic biochemical reactions acting as the engine to provide the necessary energy.

Meyerhof's published work, including the experiments he carried out during the war, did not reach England until the end of 1920.[18] Hopkins at Cambridge asked two women working in his department to replicate the experiments on frog muscle, which they did, completely confirming Meyerhof's findings. When he delivered the Herter Lectures at Johns Hopkins University in the spring of 1921, Hopkins told the audience that "any auditor would pass as satisfactory" the energy balance sheet drawn up by Meyerhof. His work clarified and enlarged Pasteur's original observation about the presence of oxygen slowing the production of the end-product of alcoholic fermentation with yeast, often referred to as the "Pasteur-Meyerhof Effect."

International Congresses of Physiologists had been held every three years in Europe from 1889. The last congress prior to the Great War had been in Groningen in the summer of 1913. It was the first that Hill had been invited to and he made quite an impression, presenting his myothermic work. The first postwar congress was scheduled to be held in Paris in 1920; invitations were sent out the previous summer to physiologists from allied and neutral countries. Meyerhof wrote to Hill complaining bitterly that he was to be excluded from the Paris congress, yet Parnas was invited.[19]

Parnas's eligibility resulted from the shifting sands of European geopolitics. He was born in 1884 in Galicia, a province of the Austro-Hungarian Empire, but in 1918 his hometown reverted to newly independent Poland. As we have seen, Parnas had been an enthusiastic supporter of German militarism in 1914, which both Hill and Meyerhof deplored. Meyerhof's main concern, however, was that Parnas would be allowed to present his erroneous findings on muscle biochemistry to an influential audience in Paris, while his own much more careful findings would go unreported. Hill decided that because German and Austrian scientists were to be excluded, he would boycott the meeting. In the event, Parnas was also prevented from attending because the Bolshevik army attacked Warsaw that summer, and he was trapped.

The Hill family moved to Manchester in March 1920, purchasing a large house in Altrincham, a commuter town. It took AV less than 15 minutes to ride to the university on Buster. His previous experience supervising the Brigands gave him the confidence to revitalize and lead a neglected department, where the number of students was swollen by the addition of more than 100 ex-servicemen. Despite his administrative and teaching duties, he did not neglect laboratory research and in this regard, his most important recruit was Mr. Downing, a talented young instrument maker. Hill was also aided greatly by Fletcher at the MRC, who dispensed not only good advice but hard cash. Fletcher wished to promote physiological research that would illuminate everyday human activity, both at work and play. He believed that the recent war had "brought into sharp relief our ignorance of the general physiology and psychology of work."[20] He recruited Hill as a member of the Council's "Industrial Fatigue Research Board" (IFRB) and encouraged the development of applied physiology at Manchester with grant monies.[21] So, in addition to investigations of twitching frog muscle using exquisitely sensitive instruments, Hill now took on the study of human motor performance. There were connections between the minimalist preparations of amphibian muscle and the muscular efforts of human volunteers: endurance, efficiency, lactic acid build-up, and oxygen consumption were common to both. Hill informed Fletcher that his ambition was to "measure, define and study the normal functional activities of man."

Hill, of course, was a noted athlete at school and had kept his trim figure by regular morning runs. He brought the competition and fun of a Blundell's sports day to the physiology department. His first co-worker, supported by the MRC, was Hartley Lupton—a Manchester physics graduate in his late twenties, who embarked on a medical degree after working at the Manchester Royal Infirmary.[22] They presented their preliminary findings to a meeting of the Physiological Society in May 1922.[23] The pair had measured levels of oxygen consumption and carbon dioxide produced, while running around an open-air track wearing a Douglas bag strapped to their backs. The runner breathed through a mouthpiece and valve system into the rubber bag that collected his expired breath for analysis. One subject was instantly recognizable to the audience: "of athletic build, 11½ stone (73 kilos), fairly fit, 35 years of age, and used to running; he is not, however, and never has been, a first-class runner." Their major finding was that at higher running speeds, the subject was using more energy than was accounted for by the oxygen supply and was going "into debt" for oxygen. At sprinting speeds, rapidly increasing oxygen debts were incurred so that "a stage is soon reached at which the accumulation of unoxidised lactic acid prevents further effort," and the runner is forced to repay the debt by consuming a huge volume of oxygen during subsequent rest. This was a canonical observation for exercise physiology and sports medicine, although more recent analyses invoke other factors such as increased temperature and steroid levels rather than just lactic acid accumulation as causing oxygen debt.[24]

Hill and Lupton were joined in 1921 by Cyril Long, a recent Manchester chemistry graduate. When Hill informed him that he needed a chemist to assay the changes in the blood of exercising animals and humans, Long was not immediately enthusiastic, but Hill soon persuaded him that there were exciting possibilities for the understanding of living processes. It soon dawned on Long, who was also a keen sportsman, that he was not being recruited just for his chemistry skills:

> When I began my work with Hill and his colleague Lupton ... I found myself running up and down stairs, or round the professor's garden while at intervals healthy samples of blood were withdrawn from my arms. When I had recovered from my exertions I was asked to sit down and analyse these for lactic acid.[25]

Hill recalled that the three of them were "rather addicted" in their "attempts to break world records of the rate of oxygen consumption during a total oxygen debt after severe exercise"—not quite the everyday social and workplace activities that Fletcher had envisaged. Yet at the limits of performance, it seemed, the human athlete and a physiological preparation of muscle converged so that the human displays "a degree of exhaustion not far short of that attained in direct artificial stimulation of the isolated muscle."[26]

Barcroft co-opted Long and Lupton to join his expedition to the Peruvian Andes in 1923 to study respiration at high altitudes. Most of the experiments employed American mining engineers, who were fully acclimatized, or indigenous males. Lupton also took part as a subject, exercising in a large glass chamber, half-filled with air and half with nitrogen. He reported Long's observations to Hill in a letter:[27]

> Shade temperature 32 deg C ... Steps timed on metronome ... 11.30 am start; 11.42 am lips bluish ... 2pm subject faint on getting up suddenly; 3.30pm finish. Lips quite blue and subject weak.[28]

AV was happy to include his family in his science endeavors. Margaret wrote to her mother describing a visit to the physiology laboratory, where she measured her lung capacity that was "more efficient than a first-class man of my weight & quite exceptional for a woman. ... On the other hand my heart is rather feeble though not diseased. ... Vivian is of course above average in both."[29] Margaret generally busied herself organizing the new house, supervising the children, buying antiques, and also found some emotional fulfillment with a new friend, Norah Clegg. Norah's father, Sir Charles Behrens, was one of the richest men in Manchester through his textile business and a governor of the university. Norah was the same age as Margaret and married to Assheton Clegg, a cotton merchant.

At the end of 1921, all four children plus AV developed scarlet fever. David developed glomerulonephritis (but less severe than Otto Meyerhof's case as a teenager), Polly had middle-ear infections, and Maurice experienced febrile seizures. AV was unable to go into the university for many weeks and their home was disinfected by the city health authority. Hill aided by Downing set up a physiology laboratory in the cellar of the house. Margaret was excited that "our cellar now is a wonderful sight with some of the most delicate apparatus ever used in physiology at work in it & a very industrious American too."[30] Wallace Fenn was a young physiologist from Harvard University on a traveling fellowship from the Rockefeller Foundation, who spent one year working with Hill (Figure 6.2).

The temperature in the basement laboratory was a steady 6–8° C; its advantage was isolation from external vibrations and electric fields such as were present in the center of Manchester. Hill suggested to Fenn that he investigate the heat produced by

Figure 6.2 Wallace Osgood Fenn (1893–1971)
Photograph courtesy of the Wellcome Collection, London.

muscle fibers allowed to shorten and lift weight, using the improved apparatus that he and Hartree had developed in Cambridge. Hill and Hartree had measured the effects during isometric contraction (where the muscle is stimulated but remains a constant length). Fenn now compared that case with allowing the sartorius muscle of frog to shorten and lift tiny weights. Fenn found that more energy (heat + work) was liberated whenever a muscle shortens versus isometric contraction. He concluded: "Not only is heat production increased as increasing loads are lifted through the same distance, but it also increases as the same weight is moved through larger distances.'[31]

The extra heat is roughly proportional to the work done. Fenn demonstrated for the first time that there was "a fairly good quantitative relationship between the heat production of muscles and the work which they perform, and that a muscle which does work liberates, *ipso facto*, an extra supply of energy which does not appear in an isometric contraction." Furthermore, the energy liberated by contraction of muscle fiber "is not dependent solely upon the initial mechanical and physiological conditions of the muscle, but can be modified by the nature of the load which the muscle discovers it must lift." Fenn was effusive in his thanks to Prof. A.V. Hill, saying that he instructed him in technique and gave invaluable assistance throughout and he paid Margaret "grateful acknowledgements" too. AV heralded the important finding as the "Fenn effect," as it is still referred to nearly one century later.[32] The Fenns were especially indebted to Margaret for organizing accommodation and helping with their baby son.

Otto Meyerhof made his first postwar trip to England in the summer of 1922, crossing the North Sea to Harwich and then by train to Manchester. During the course of his visit, he politely pointed out to Hill that he was suspicious of research work being published by Dr. J. Holker from the Manchester pathology department. Embarrassingly, Hill had just forwarded one of Holker's papers to the Royal Society for publication. Meyerhof was the only skeptic—Professor Dean, the editors of three scientific journals, colleagues in the pathology and physiology departments, and Hill had all been taken in. Meyerhof stayed in Manchester for one week before Hill transported him in Buster's sidecar to Cambridge. During the journey, a bolt holding the sidecar to the bike sheared and had to be "forged anew by a blacksmith"[33] en route.

Alerted by Meyerhof, Hill became increasingly suspicious about Holker's work. Holker was supported by an MRC grant and by March 1923, Walter Fletcher started to question whether it should be continued. Hill decided to ask Holker if he could witness some experiments, and he gladly agreed. "On no occasion," Hill observed, "were results obtained similar to those previously reported and published."[34] He informed the MRC. Holker wrote a letter to Fletcher, claiming that the presence of a witness had introduced a disturbing psychological factor. Fletcher replied that "is the claim commonly made by spiritualists" before ending the grant. Hill was most puzzled by Holker's lack of "resentment, or distress, or indignation, at … a direct challenge to his integrity." Holker withdrew from the university and opened a venereal diseases clinic in Manchester.

As mentioned earlier, Ernest Starling stayed with the Hills for the November meeting of the Physiological Society, and during those few days decided that AV should succeed him as Jodrell Professor of Physiology at University College London. The main political issue to be settled at the Manchester meeting was the question of the next International Congress due to be held in Edinburgh in July 1923. The president of the Congress and Edinburgh professor, Sir Edward Sharpey-Schafer, had lost both his sons in the war. He was determined to avoid the restriction on former enemies that had marked the Paris conference and to invite all eligible physiologists. He was alarmed by rumors that the French were threatening a boycott if German scientists were allowed to attend. The senior British physiologists were steadfast in their support for Sharpey-Schafer, and invitations were issued to all.

In January 1923, the French Army of the Rhine marched into the Ruhr after Germany had consistently failed to make scheduled reparations.[35] Animosity between the two nations was unbridled and the German economy, deprived of its industrial heartland, collapsed. Hyperinflation quickly followed, so that by the time of the July congress, the exchange rate for the mark against the dollar had fallen from less than 8,000 at the beginning of 1923 to one million. Only about three French physiologists came to Edinburgh, but many Germans managed to attend. Hill felt they were grateful to be treated like civilized people again.[36] Hill and his young associates, Fenn, Long, and Lupton, all presented results of their muscle research, but the scientific limelight belonged to the Scottish-born professor of physiology from the University of Toronto, John J.R. Macleod, in whose laboratory purified insulin had been extracted from the pancreas of dogs some months earlier. His junior colleague, Frederick Banting MD, was also present.

There is no doubt the German contingent enjoyed the hospitality and the opportunity to exchange ideas again in a relaxed environment. They immediately invited Hill, Starling, and Edward Cathcart, a medically qualified biochemist from Glasgow who had spent time working in Germany, to come to their national physiology congress in Tübingen in early September. They were joined by Herbert Gasser, a neurophysiologist from St. Louis, who was on an extended European tour funded by the Rockefeller Foundation. After the congress, Hill and Meyerhof went hiking in the hills around the origin of the Danube river for several days. Hill said they did discuss the neutralization of the Ruhr-Rhineland briefly, but spent more time talking about the neutralization of lactic acid in frog's muscle.[37]

Notes

1. Fletcher and Hopkins (1917).
2. A.V. Hill to W.M. Fletcher (14/12/1916) AVHL II 4/57.
3. A.V. Hill to J. Barcroft (20/8/19) AVHL I 3/1.
4. J. Barcroft to A.V. Hill (21/8/19) AVHL I 3/1.

5. F.J.W. Roughton, "Joseph Barcroft (1872–1947)," *Obituary Notes of Fellows of the Royal Society* 6, no. 18 (1949): 315–345.

6. A.V. Hill to W.M. Fletcher (12/5/19) AVHL II 4/57.

7. A.V. Hill, "William Hartree (1870–1943)," in *The Ethical Dilemma of Science* (1960), 173–178.

8. A.V. Hill and W. Hartree, "The Four Phases of Heat-Production of Muscle," *Journal of Physiology* 54 (1920): 84–128.

9. A.V. Hill, "William Stirling, Showman" M&R 477–480.

10. Note in AVHL II 5/36/19.

11. W.L. Bragg to A.V. Hill (5/11/19) AVHL I 4/11.

12. A.V. Hill to W.M. Fletcher (15/12/19) AVHL II 4/57.

13. Hill, "William Stirling."

14. Mark Jones, "Kiel Mutiny, 1914–1918," in *International Encyclopedia of the First World War*, ed. U. Daniel, P. Gatrell, O. Janz, et al. Issued by Freie Universität Berlin, Berlin 2016-05-19. DOI: 10.15463/ie1418.10908 (accessed 30/12/19).

15. W. Meyerhof (2002), 15.

16. O. Meyerhof, "Über die Energieumwandlungen in Muskel," *Pflüger's Archiv für die gesamte Physiologie* 182 (1920): 284, quoted in Teich (1992), 220–221.

17. Rall (2014), 5.

18. J. Needham, "Sir Frederick Gowland Hopkins OM FRS (1861–1947) Centenary Lecture," royalsocietypublishing.org/doi/10.1098/rsnr.1962.0014 (accessed 30/12/19).

19. A.V. Hill, *Trails and Trials in Physiology* (London: Edward Arnold, 1965).

20. W.M. Fletcher, "The Growing Opportunities of Medicine" (1926) copy in AVHL II 4/27.

21. M. Stadler, "Assembling Life: Models, the Cell, and the Reformations of Biological Science, 1920–1960" (PhD thesis, Imperial College, University of London, 2009).

22. A.V. Hill, "Hartley Lupton (1893–1924)." in Hill (1960), 142.

23. A.V. Hill and H. Lupton, "The Oxygen Consumption during Running," *Journal of Physiology* 56 (1922): xxxii–xxxiii.

24. D.R. Bassett, "Scientific Contributions of A.V. Hill: Exercise Physiology Pioneer." *Journal of Applied Physiology* 93 (2002): 1567–1582.

25. O.L.K. Smith and J.D. Hardy, "Cyril Norman Hugh Long (1901–70)," *National Academy of Science Biographical Memoir* (1975): 265–309. Hill took Long to UCL with him and in 1925 arranged for him to complete an MD at McGill University. Subsequently he was chairman of the biochemistry and physiology departments at Yale for over thirty years.

26. A.V. Hill, C.N.H. Long, and H. Lupton, "Muscular Exercise, Lactic Acid, and the Supply and Utilization of Oxygen," *Proceedings of the Royal Society of London (B)* 96 (1924): I–III 438–475; IV–VI 97: 84–138.

27. H. Lupton to A.V. Hill (6/7/23) AVHL II 4/56.

28. In July 1924, Lupton developed a large thoracic tumor with obstruction of his superior vena cava. His mother sent regular updates on his condition, and when he died in September, she thanked Hill "for all the happy moments you gave to him." In his obituary for *Nature*, Hill (1960, 142–143) painted him as a pioneer in the field of exercise physiology and said, "he was never so happy as when 'going all out.' "

29. Angier I (1978):165.

30. Angier I (1978):174.

31. W.O. Fenn, "A Quantitative Comparison between the Energy Liberated and the Work Performed by the Isolated Sartorius Muscle of the Frog," *Journal of Physiology* 58 (1923): 175–203.

32. Wallace Fenn (1893–1971) After two years in England, he became chairman of the department of physiology at the University of Rochester. Later in his career he made contributions to human respiratory physiology and was an expert advising NASA in the 1960s.

33. O.F. Meyerhof to A.V. Hill (16/12/49) AVHL I 3/58.

34. A.V. Hill, "The Piltdown Skull in My Cupboard" M&R: 484–488.

35. A. Tooze, *The Deluge* (London: Penguin, 2015), 442–445.

36. A.V. Hill (1969).

37. www.nobelprize.org/prizes/medicine/1922/hill/speech/ (accessed 21/1/20).

7

Stockholm and Beyond

In his will, Alfred Nobel stipulated that there would be monetary prizes awarded annually to those who have conferred the greatest benefit to humankind. One of the five areas to be recognized was "the most important discovery within the domain of physiology or medicine." He directed that responsibility for choosing recipients would rest with a committee at the Royal Caroline Medico-Surgical Institute, now known as the Karolinska Institute (KI), in Stockholm. The institute was less than a century old, having been founded by the king in 1810 to improve the training of army doctors. It was a less venerated medical school than those at the universities of Uppsala or Lund, and Nobel's choice surprised the medical establishment. One reason he picked it was because from its founding, the institute promoted scientific knowledge as the basis for the practice of medicine.[1] An early professor at the Royal Caroline Medico-Surgical Institute was Jöns Jacob Berzelius, a founder of modern chemistry, who discovered several elements and was the first to suggest that catalytic reactions are essential to the chemistry of living systems.[2] It was also Berzelius in 1847 who first identified the acid in the fluid extracted from the muscles of hunted stags as identical to that from sour milk.[3]

A few years before he died in 1896, Nobel invited a young physiologist from the KI, Johan Erik Johansson, to spend some months in his laboratory outside Paris, to improve the techniques for blood transfusion. Working with Johansson cemented Nobel's intention to make the KI responsible for administering the physiology or medicine prize. By the time of Nobel's death, Johansson was a professor of physiology at the KI and became a leading voice in interpreting the terms of the will. A Nobel Assembly of fifty faculty members at the KI, as well as well-known medical scientists in other countries, could nominate candidates: the final recommendations would be made by a Nobel Committee of five people. The number of nominations in the early years of the twentieth century varied between 80 to 150 annually, and the medicine or physiology prize was awarded to one or two laureates every year from 1901 to 1914. Then there was a four-year hiatus due to the Great War before the prize was reinstated for 1919 and 1920. No medicine or physiology prize was awarded again in 1921, nor in 1922, and the prize monies were rolled over in a special fund.

The prize money each year was the interest on government bonds that Nobel had arranged through his will. With the disruption of the war and subsequent economic downturn, the prize money in the early years after the war was only about half what it had been prior.[4] The nominations for 1921 were headed by the Oxford

Bound by Muscle. Andrew Brown, Oxford University Press. © Oxford University Press 2023.
DOI: 10.1093/oso/9780197582633.003.0007

neurophysiologist Sir Charles Sherrington, who had been proposed nearly every year since 1902. He was popular in the list again for 1922, when half a dozen candidates underwent full review before the prize was withheld. In 1923, the Nobel Assembly decided that they would award the 1922 prize retroactively as well as the current one. For the first time, Albrecht Bethe, Meyerhof's prewar chief at Kiel, was asked for his recommendations and he proved to be an enthusiastic participant.[5] He submitted one dozen names in four groups of three; while the logic of these groupings may have been obvious to Bethe, it was not necessarily clear to the Nobel Assembly, which had not to that time ever divided the physiology prize three ways. Hill was linked with Gustav Embden and with Bethe's friend and successor at Kiel, Rudolf Höber. Another triad included Meyerhof, and two other biochemists, Leonor Michaelis and Ernest Overton. The Nobel assembly decided that the theme for the 1922 prize would be muscle physiology, where in the estimation of Professor Johansson, developments had been "colossal." The Nobel Committee selected three final candidates to have detailed reviews: Hill, Meyerhof, and Embden.

Gustav Embden was about one decade older than the other two nominees. He was professor of physiology in Frankfurt-on-Main, and, like Meyerhof, was essentially a biochemist. Before the war, he had explored the similarities between alcohol fermentation and lactic acid formation in muscle, seeking to establish whether as in the fermentation of yeast, lactic acid could be formed from carbohydrate in a cell-free preparation of muscle juice. He found that it could, so he then wondered about the role of phosphate that Harden and Young had identified in their work with yeast. In 1914, he reported with co-workers that the formation of lactic acid in muscle press-juice is accompanied by phosphoric acid formation. During the war, Embden tested the physical performance of soldiers and miners and found that it was improved by the addition of phosphate to the diet.[6] In 1921, Meyerhof, using chopped muscle, became convinced of the centrality of phosphate to lactic acid production. When he suspended the muscle pieces in solution buffered with phosphate, the amount of lactic acid produced was greatly increased and no other buffer had this effect.

Johansson took on the scrutiny of Hill and Meyerhof himself. John Siqvist, professor of chemistry and pharmacology, evaluated Embden's biochemical work and judged it "to be highly deserving of a Nobel Prize in Physiology or Medicine."[7] As chairman, Johansson encapsulated the choice that faced the committee:

> If the three researchers had worked together, the formulation for awarding the prize this year would be very simple: a prize for the discovery of the physical-chemical processes behind the muscle process should be awarded jointly to professors Embden, Hill and Meyerhof, who together have made this discovery. However, the situation is complicated by the fact that the three researchers have worked independently, and only the general flow of thoughts and in part the research methodology have been shared. In fact, it would be difficult to imagine

that such widely encompassing and, as regards resources, highly demanding work, could have been concentrated in a single place. The question one can then pose is: are the separate contributions, which each one of the three researchers has made to the joint effort, of such a nature that they can be viewed as representing single discovery, and can one attribute to this such a degree of importance that it is necessary to award a Nobel Prize. ... I find it obvious that each one of these discoveries has the importance that the statutes require for the awarding of a prize. Like the proposer, Bethe, I find it impossible to put any of these contributions ahead of the other.[8]

Johansson's summary first establishes, in rather convoluted fashion, that the researches of the three scientists can be considered as one body of work; then he concludes that there is nothing to choose between the inputs of the three candidates, all of whom have made contributions worthy of a Nobel Prize. There is no record of the committee's final deliberations but, at a faculty meeting of the Karolinska Institute on 11 October 1923, there was a unanimous recommendation that the prize for 1922 should be shared equally by Hill and Meyerhof.[9] There is no further mention of Embden, who would be nominated unsuccessfully eight more times for the physiology or medicine prize and once for the chemistry prize, over the next decade. Although the statutes for the medicine prize explicitly allow the prize to be shared by two or three workers (but not more than three), it was not divided between three scientists until 1934.

A few days after Margaret Hill read the announcement in *The Times* of the 1922 Nobel prizewinners, AV received a letter in the physiology department at UCL from Otto Meyerhof.[10] It was typewritten in rather formal German and expressed his heartfelt happiness both in winning the prize and sharing it with Hill. Like most laureates down the years, he was also surprised. Meyerhof was naturally already thinking about his Nobel Lecture to be delivered in six weeks' time; he wanted his and Hill's to complement each other. He suggested that Hill should go first to talk about his myothermic research, without straying into the underlying biochemical reactions that he would cover. He was nervous that there was so much to do before they met again in Stockholm.

Margaret decided not to accompany AV to Stockholm and stayed at home with the children. Hedwig Meyerhof did come, although she was still recovering from severe postpartum depression following the birth of her third child, Walter, in April 1922.[11] Her portrait was featured on the cover of *IDUN*, a Swedish women's magazine, which described her as shy and "a soulful beauty with intelligence and gentle motherhood." The prizes, according to tradition, were presented on 10 December, the anniversary of Nobel's death (Figure 7.1).

Professor Johansson as chairman of the Nobel Committee of the Royal Caroline Institute made the introductory speech at the award ceremony.[12] It was a masterly summary, beginning with the progressive triumph of physiology over vitalism,

Figure 7.1 The Nobel Prize ceremony. Hill sitting with Meyerhof to his right, and W.B. Yeats to his left

Photograph courtesy of Professor Nick Humphrey/Churchill Archive Centre, Cambridge (Papers of A.V. Hill, AVHL).

touching on the application of thermodynamics to muscle activity, before coming to the work of the two laureates. He summarized their achievements, as follows:

> Hill has analyzed, by means of an extremely elegant thermoelectrical method, the time relations of the heat production of the muscle; and Meyerhof has investigated by chemical methods the oxygen consumption by the muscle and the conversion of carbohydrates and lactic acid in the muscle.

After setting out their research findings in more detail, relating both to the previous contributions of Fletcher and Hopkins, Johansson came to his peroration, before presenting the men to the king.

> Professors Hill and Meyerhof. Your brilliant discoveries concerning the vital phenomena of muscles supplement each other in a most happy manner. It has given a special satisfaction to be able to reward these two series of discoveries at the same time, since it gives a clear expression of one of the ideas upon which the will of Alfred Nobel was founded, that is, the conception that the greatest cultural advances are independent of the splitting-up of mankind into contending

nations. I also feel confident that you will be glad to know that the proposition which has led to this award of the Nobel Prize originated from a German scientist [Bethe] who, in spite of all difficulties and disasters, has clearly recognized the main object of Alfred Nobel.

That evening was the prizewinners' banquet at the Grand Hotel where, as AV told Margaret, he mixed with Swedish royalty, ten members of the Nobel family, a poet, and about one hundred professors. The poet was W.B. Yeats, who won the 1923 Prize for Literature. Hill took an instant dislike to him, perhaps because Yeats kept insisting that his award was not for his own poetry, but given in recognition of Irish literature and to celebrate the recent independence of Ireland.[13] Hill wrote to Margaret that Yeats is "a comic-looking bloke,"[14] warning her—probably in jest—that if Yeats talked about the wrongs of Ireland, he would counter with the wrongs of England. In fact, Hill delivered a humble speech, extolling the intrinsic reward and adventure of scientific research. He did manage to slip in a sly dig that scientists, unlike poets, encountered the romance of discovery on a daily basis. He echoed Johansson's morning remarks about the internationality of science that should be blind to race, creed, age, and status, and rejoiced that physiology was the first subject to revive a truly international congress after the war. Referring to Meyerhof as his colleague and friend, Hill said that they both viewed the joint honor as "a seal of your approval, the approval of a people friendly to my nation and to his nation. Of a brotherhood in Science between a German and an Englishman."[15]

Speaking in German, Meyerhof expressed his deep appreciation to his Swedish hosts and stated that the award to him and his friend, Hill, was unequivocal evidence that science must set aside national barriers and prejudices.[16] Revealing his poetic leanings, he continued that the prize was "a reminder to all representatives of science to pour onto the excited waves of the national passions of the peoples, the smoothing oil of genuine wisdom and national self-contemplation. ... In this sense we can consider all Nobel Prizes as 'Peace Prizes.'" However many million German marks Meyerhof received for his half share, Hill banked 3,664 pounds and threepence (which represented a major chunk of the cost of his new house in Highgate).

The two laureates delivered their Nobel Lectures two days later. As suggested by Meyerhof in his letter, Hill went first and concentrated on his myothermic experiments. He made the point that the changes during muscle activity take place with great rapidity, and although, fundamentally, they depended on biochemical processes, chemical investigation was a slow and laborious business. Despite the technical difficulties of his temperature recordings during muscle stimulation, the law of the conservation of energy provided a firm theoretical base so that the "advantage of the study of the thermodynamics of muscle is that heat may be measured in absolute units, rapidly and at once, and the time-course of its evolution analyzed by suitable means."[17] Once physical investigations such as his had sketched the outline of the mechanism of muscular contraction, he believed it would be possible finally to "have the complete chemical picture painted into it." During his lecture, Hill paid tribute to the work of many of his friends and colleagues, past and present.

Hill referred to some work he had conducted on male and female students at Manchester in which, unusually, two of the subjects studied were children—their ages matched Polly's and David's. The subjects rotated a heavy flywheel by pulling on a string rigged through a variety of pulleys. Hill observed that the force exerted by a muscle decreased as the speed of shortening increased.[18] He explained to the audience in Stockholm:

> In the case of human-arm muscles the work does not attain anything near the maximum value unless the contraction has been opposed by a mass large enough to make it occupy at least two seconds.

There had been theories dating back fifty years that a stimulated muscle was like a stretched spring, where the stored tension could be converted into work or heat as the elastic muscle shortens. Yet in the case of a simple elastic body such as a spring, the force generated depends only on its length and not the speed of contraction. Hill modified the concept by supposing during muscle contraction that energy is expended in overcoming the viscous resistance of the muscle to a change of shape: the faster the muscle contracted, the more energy would be needed to squeeze the viscous fluid in the fibers into a new form. This became known as the "visco-elastic theory." By the time of his Nobel Lecture, Hill was already having doubts about its validity. Wallace Fenn's results were incompatible with such a theory, and Hill was fully aware of Fenn's paper,[19] then awaiting publication that stated with authority:

> The shortening of a muscle appears to be an active process and not merely analogous to the release of a spring previously stretched. The energy used in the performance of work is developed at the time when the work is done and does not represent potential energy already developed before shortening begins.

Despite this, Hill remained on the fence over the visco-elastic theory for several more years. When he delivered his own Croonian Lecture to the Royal Society in 1926 on "The Laws of Muscular Motion," he still seemed in thrall to "the analogy of an elastic system possessing considerable viscosity."[20] Reminiscing over forty years later, Hill wrote that Fenn's conclusions "were obviously the death warrant of the visco-elastic theory; yet I went on thinking in terms of it for a long time after. It is odd how one's brain fails to work properly when pet theories are involved."[21]

Meyerhof, in his Nobel Lecture on "Energy Conversions in Muscle," acknowledged the pioneering work of Fletcher and Hopkins, before referring to Hill's elucidation of the heat produced on muscle contraction and during subsequent recovery as "shining out like a beacon light through a sea mist, [making] it possible for me to steer a safe course through the shallows."[22] In his discussion of the chemical reactions underlying muscle activity, Meyerhof also took a thermodynamic approach and presented estimated values of the combustion heats of glycogen and lactic acid. He emphasized that his results were in keeping with Hill's findings.

Meyerhof was the leading proponent of an alternative to the visco-elastic theory of muscle contraction. He believed that lactic acid not only represented part of the fuel cycle for muscle contraction in its formation from glycogen, but it was also part of the machinery of muscle shortening. He postulated that in its resting state, the protein responsible for shortening was in an ionized state, whereas in the presence of lactic acid, it would become un-ionized and undergo a change in conformation: contraction of the muscle fibrils. Meyerhof touched on the lactic acid theory toward the end of his Nobel Lecture:

> We can, however, say now that the deionization of the protein by the lactic acid produced by muscle activity plays, without any doubt, an important part in the contraction mechanism. It explains in the first place the flaccidity of the muscle which sets in after shortening under anaerobic conditions in spite of the presence of lactic acid. This flaccidity would be brought about by nothing more than the diminution of the lactic acid acidity, just as, conversely, we would hold the H-ions responsible for the release of the contraction.

Just as with the visco-elastic theory, Meyerhof's lactic acid contraction idea would face incontrovertible evidence to the contrary over the coming decade. Ironically, Gustav Embden delivered the first blow in 1926, when he demonstrated that under anaerobic conditions, some of the lactic acid is formed after contraction.[23]

In many ways, Hill had been Meyerhof's closest colleague since the war. Mostly he had worked alone in a small, poorly equipped laboratory in Kiel until acquiring an unpaid assistant for the first six months of 1922, Hans Hermann Weber. He was a newly minted MD already interested in the chemistry of muscle, who would go on to enjoy a distinguished career in the field. Just prior to the announcement of the Nobel Prize, Professor Höber had finally succeeded in persuading Kiel University to create a chair in physiological chemistry. Although he lobbied vigorously for Meyerhof to have the position, the university appointed a marine biologist, August Pütter. The decision was influenced by the fact that both Höber and Meyerhof were Jews.[24]

Meyerhof made a tour of the United States in the spring of 1923, traveling by train and giving lectures at all the major medical schools east of the Mississippi. The lectures were well received and Simon Flexner, the first director of the Rockefeller Institute and Abraham Flexner's brother, offered him a position in New York. The city made a "fabulous impression"[25] with its busy traffic and countless skyscrapers under construction, but Meyerhof was not ready to leave Germany. He told Warburg that as a family man he did not think he would find a better way of life in the United States but, on the plus side, living costs were not as high as he expected. Later in the year, even with the Nobel Prize in his pocket, Meyerhof's applications to German universities were repeatedly rejected. Sensing that Meyerhof's talents might be lost to the nation, his mentor, Otto Warburg, now head of the biology institute of the Kaiser

Figure 7.2 Meyerhof's research team at the KWI, Berlin-Dahlem. Left to right, Karl
Lohmann, Otto Meyerhof, Ken Iwasaki, David Nachmansohn, and Alexander von Muralt
Photograph courtesy of Marianne Emerson and the Meyerhof family.

Wilhelm Society (Kaiser-Wilhelm-Geselschaft, KWG) in Berlin-Dahlem, began to
agitate for Meyerhof to be put in charge of a physiology department there.[26] Warburg
convinced several colleagues to give up laboratory space, and Meyerhof moved to the
KWG complex in 1924. His department was split between different floors in separate buildings, but for the first time as an established researcher, he had ready access
to a group of top scientists (Figure 7.2). His first hire proved to be inspirational: Karl
Lohmann, a man in his mid-twenties from a farming background, who had served
in the German artillery on the Western Front in the war and had just obtained a doctorate in chemistry.

Hill made his own first visit to America in October 1924. He was away for nearly two months. Margaret took advantage of his absence to invite Eglantyne Jebb to stay for a few days. She was between a stint in Geneva, promoting the Declaration of the Rights of the Child at the League of Nations, and an arduous tour of Europe to persuade political leaders to sign the charter. Margaret worried that she looked frail and would catch chickenpox from her children.[27]

The centerpiece of AV's trip was a series of six Herter Lectures at Johns Hopkins Medical School, soon published in book form as *Muscular Activity*. On his return to England, Hill gave an address to the BAAS on the physiological basis of athletic records. He presented data on records in running, swimming, and rowing, in graphical form that were the forerunners of curves that plot human power output over time. He explained the speed difference between sprints and longer races in terms of oxygen-dependent and oxygen-independent systems.[28] His seminal ideas about exercise physiology and biomechanics reached a wider audience when he had a short piece published in *Scientific American*.[29] It was accompanied by an eye-catching front cover to the magazine, suggesting that the human body could be regarded as a machine whose energy expenditure is capable of accurate measurement.

1926 was a momentous year for Hill. The Royal Society offered him a research professorship, the Foulerton chair, to be held at UCL. This appointment relieved him of teaching and administrative responsibilities, and he would hold the chair for the next quarter of a century. Mr. Downing continued to make improvements to the laboratory equipment and insulated the cellar, as far as possible, from electromagnetic and vibrational interference. A three-ton concrete pillar extended through the subsoil of London and was surmounted by a heavy glass platform loaded with iron weights and floating on a bath of mercury: the sensitive recording equipment sat on top of the platform. Magnetic fields were reduced by steel shielding and wrapping objects in special alloy tape. The cellar was rewired inside earthed steel tubes. Hill ruefully observed that these precautions would have proved adequate working anywhere but in a great city.

Experiments were carried out at night and early on Sunday mornings, including Easter, when the city was quiet. Hill with Downing and a visiting post-doctoral student from Chicago, Ralph Gerard,[30] gathered to observe the minuscule heat produced during the passage of an impulse along sciatic nerve dissected from a pithed frog. Each nerve specimen tested weighed about 50 mg so that the biological material at the heart of the experiment was dwarfed by the gargantuan apparatus used to examine it. Since Helmholtz, many physiologists (including Hill prior to the war) had attempted unsuccessfully to measure heat associated with nerve activity. In the quiet, heavily shielded laboratory, Downing's latest thermopiles and galvanometers registered an "excessively small" heat measurement of 0.000069 cal per g per second. Gerard spent long winter's nights, when there was a minimum of outside interference, making readings from the galvanometer every five seconds.[31] Hill had already structured the final paper and they just had to add the results that made him conclude "that the process of transmission does *not* involve the whole surface of the nerve fiber, but only a small proportion of it."[32]

Hill and Gerard presented the results of nerve heat production at the International Congress of Physiology in Stockholm in the summer of 1926. Following the congress, Gerard traveled to Berlin with Meyerhof to begin a year working in his laboratory. On arrival he found that Meyerhof had already assembled apparatus for him to measure the oxygen consumption of frog nerve. Gerard quickly showed that the respiration quantitatively matched the heat production.

As a result of his deepening friendship with Meyerhof and impressed by the advances being made by Hopkins (now the first professor of biochemistry at Cambridge), Hill hired a biochemist for his laboratory UCL. Philip Eggleton, a man in his mid-twenties, tall with curly hair, had no previous academic experience when he arrived at UCL in 1925. The following year, a recently qualified physician, Grace Palmer, who had already obtained a first-class degree in physiology at UCL, arrived in Hill's laboratory to start her own research career. She and Philip worked together and soon married (Figure 7.3).

Figure 7.3 Philip Eggleton and Grace Palmer Eggleton
Photograph courtesy of the Wellcome Collection, London.

They submitted their first paper on New Year's Eve 1926, describing the presence of an organic phosphate in muscle that was very unstable in acid solution.[33] They proposed the name *phosphagen* for the compound, whose concentration decreased with muscle stimulation and recovered during subsequent rest. They found phosphagen in all vertebrate muscles (except the involuntary smooth muscles and heart), but not in the muscles of invertebrates. A few months later, the Harvard laboratory of Professor Cyrus Fiske reported that phosphagen was a compound of creatine (an amino-acid-like chemical, found in muscle), more accurately *phosphocreatine*. Meyerhof and Lohmann soon reported that the phosphocreatine represents a tremendous reservoir of energy; when it is hydrolyzed with the liberation of inorganic phosphate, about 60 calories of heat are released per gram of the substance.[34]

These developments were seized on by Meyerhof in Berlin. When he interviewed David Nachmansohn, a newly qualified physician, they talked for an hour about muscle physiology and the newly discovered phosphocreatine. Nachmansohn had embarked upon a career in biochemistry, taking a research post in the pathology department of a large Berlin hospital, where he became friendly and worked with H.H. Weber, Meyerhof's former assistant in Kiel.[35] Weber suggested to him that he too should spend time working with Meyerhof. Meyerhof may have recognized something of himself in the young man, who was raised in a liberal Jewish family, educated at a humanistic gymnasium, and was a devotee of poetry, especially Goethe's. After saying, "Usually I don't accept beginners," Meyerhof agreed to accept Nachmansohn into his small department. He asked him to explore connections between phosphocreatine breakdown, lactic acid formation, and muscle contraction under anerobic conditions. Nachmansohn made wide-ranging observations. Unlike many biochemists, he also considered muscle histology, noticing that rapidly contracting "white" muscle fibers contained much higher concentrations of phosphocreatine than slowly moving "red" muscle. Nachmansohn flourished in the rarefied atmosphere of the KWG, which housed many of the giants of German science, and he found his regular discussions with Meyerhof especially inspirational.

In November 1926 the Royal Society awarded Hill its Royal Medal for biological science, with Sir Ernest Rutherford PRS describing his research on muscle and nerve physiology as "a success beyond expectation."[36] Hill celebrated the Christmas season by delivering a wildly successful series of lectures and demonstrations for children at the Royal Institution on "Our Nerves and Muscles: How We Feel and Move." When AV told his children what he was preparing, Janet, aged eight, insisted on taking part and her siblings joined in enthusiastically (Figure 7.4).

As Margaret noted, this caused "a great newspaper stunt." The *Daily Mail* reported on the first talk, "Nerves and the Messages They Carry," and described how David, aged twelve, had his ulnar nerve electrically stimulated making his little finger flex—"until the little finger oscillated thirty times a second, far faster than it could move without electrical stimulation. ... The next subject was David's sister Polly. She put her hand to a wire through which a current of 500,000 volts was passing. Polly only smiled."[37] Janet had her heart sounds amplified as she ran up and down the steep auditorium

Figure 7.4 AV and the Hill children preparing the Royal Institution Christmas Lectures for 1926. (left to right): Polly, AV, Maurice, Janet, and David

Photograph courtesy of Professor Nick Humphrey/Churchill Archive Centre, Cambridge (Papers of A.V. Hill, AVHL).

stairs, and the movements of her stomach were displayed by X-ray fluoroscopy. In the last lecture, "Speed, Strength and Endurance," David showed more determination than perhaps his father had bargained for; he rode on a stationary bicycle, breathing through a mask supplying him with air containing only 13 percent oxygen. According to the *Sunday Times*, "David declined to admit a proper tiredness within the expected time-limit. Those near him could see that the colour ebbed from his cheeks ... stealing an occasional sly glance at his father, he still pedalled on." AV called for "super air" with 80 percent oxygen to be substituted for the mountain-top variety so that David quickly recovered and showed off with a sprint finish. Some months later, Hill received a letter from one of the British anti-vivisection societies citing his "abuse of parental authority" as an example of how "physiologists regard all living beings (including their own relatives), as mere material for experimentation."[38] The supposed maltreatment of the children became a running family joke. Hill published the lectures in book form as *Living Machinery*. He dedicated it to his "four assistants and experimental subjects" thanking them "for their unfailing zeal and co-operation."[39]

Fleet Street's interest in the Christmas lectures was more welcome than press coverage of a criminal trial in November. A tradesman, who supplied animals and food to UCL, was convicted of stealing two dogs, which he said he intended to sell to the physiology department.[40] In the early twentieth century, there was a sustained expansion in the use of living, vertebrate animals for medical research and teaching, with physiologists being the main exponents. The new head of department, Charles Lovatt Evans, was left to explain that he accepted only animals that came with a certificate of ownership and that the treatment of laboratory animals followed the rules set by the Home Office. Professor Hill may not have helped his attempts to calm the public by writing a letter to *The Times*[41] pointing out that 50,000 stray dogs were being euthanized by the authorities every year just in London; because of the anti-vivisectionist inspired Dog's Act of 1906, none of them could be used for experimental purposes. Hill stated that if just 1 percent of the dogs were used for physiology experiments, there would be no need for UCL to buy stolen ones.

Embarrassingly for UCL, within weeks it was at the center of another dog scandal, when an elderly gentleman came straight there after his Irish wolfhound went missing. A quick check of the cages in the animal house revealed his pet, luckily unharmed. The incident led the British Union for the Abolition of Vivisection to bring a case, unsuccessfully, against the professor of pharmacology, Ernest Verney. The medical profession and physiologists had their own effective pressure group—the Research Defence Society. In 1912, Hill became the honorary secretary of the Cambridge branch, and he remained a stout opponent of anti-vivisectionists. He received some threatening letters and phone calls in the wake of his letter to *The Times*. He decided to include details of these in a flippant essay on anti-vivisection he wrote for *The Nation*, a political weekly partly owned and overseen by J.M. Keynes. His article was too much for A.A. Milne, who had just risen to fame as the author of *Winnie the Pooh*. Milne wrote a long letter, published under the heading "Vivisection and the Expert."[42] After stating he was rushing in where "others have already been trodden on" to engage "the triumphant Professor Hill," he set out his qualifications as a non-expert so that Hill could not expose them later. He believed that the world was not made entirely for experts. In his view, the only question requiring an expert answer was the scientific value of the experiments on live animals (where he would take Hill's word), but he doubted whether the value justified the cruelty. He assumed that if Hill did reflect on the ethics of the practice, he would probably agree that some limit should be drawn. Milne continued:

> Indeed from his last letter it would seem that he himself draws it, indignantly, at the vivisection of criminals. Well, that is something. ... If Professor Hill can imagine some giant personified vivisection starting operations with a fly, and working relentlessly upwards, through rabbits, dogs, literary men, horses, criminals, ordinary men, women and children to the extreme height of the scientist, he would see that at every stage there would come from mankind an increasing volume of protest. ... It is not, then, not for experts but for the whole community to decide when "decency forbids." Does it forbid the vivisection of dogs? Many have decided it does; many have decided it doesn't. Some of us are still a little uncertain.

Milne's letter was published on 15 January 1927. Hill did not see it because he had already left for a long sabbatical in the United States with Margaret and Polly. He spent several months as a visiting lecturer at Cornell University and was mainly occupied in novel experiments with the college track team. He devised a method for measuring a sprinter's acceleration and velocity by having them wear a magnetic band around their chests. They ran along a track, where he had positioned coils of wire at known intervals. As the runner passed each coil, a pulse would be generated and recorded by a galvanometer. Knowing the distance between each coil and the elapsed time, the runner's velocity could easily be calculated for successive stages of the run (Figure 7.5).

After AV's return to England in June, the family went on a camping holiday on Dartmoor. Hill had spent the previous summer at the Marine Biological Station (MBS) in Plymouth, "to learn some Zoology." In the summer of 1927, Meyerhof was also working at the MBS and he joined the Hills camping. During their stay, the Hills were looking for a suitable holiday home: AV loved Devon and the MBS offered him the perfect summer diversion. While Meyerhof was with them, they saw a modest house standing in a triangular field. They bought it from the farmer who built it, and he agreed to add some modern conveniences such as bathrooms. The house, Three Corners, was less than 10 miles from Plymouth and would provide "a fairyland for family and many guests."[43] It would also become a social annex for the MBS. The following summer, members of the national and West Country press descended on Three Corners after F.G. Donnan, a professor of chemistry at UCL, had given a lecture at the annual meeting of the British Association in which he made some hyperbolic remarks

Figure 7.5 Hill measuring a sprinter's acceleration at Cornell University, 1927
Photograph courtesy of Professor Nick Humphrey/Churchill Archive Centre, Cambridge (Papers of A.V. Hill, AVHL).

about his colleague being on the eve of "a discovery of astounding importance."[44] This produced sensational newspaper headlines such as "Man Is the Creator of Life," which Hill found out about as he was spending the day at the MBS. After returning from an evening swim with his children and their numerous friends who were camping in the garden, AV had to persuade different journalists that it was all nonsense.

While Hill was taking an interest in the biophysics of muscle activity at the scale of the human individual (he was an interested observer of the 1928 Olympic Games in Amsterdam), Meyerhof was fully engaged in uncovering the sequence of chemical reactions that constitute the breakdown of glucose, or glycolysis, at the molecular level. He used a system that simply consisted of crushed frog or rabbit muscle dissolved in potassium chloride at −1° C. He found that a cell-free filtrate was just as productive as chopped muscle in producing lactic acid from a number of carbohydrates.[45] Back in 1912, Hill had suggested that there was time just for a single biochemical reaction to transform the precursor molecule of glucose into lactic acid, but Meyerhof had come to realize that glycolysis comprised a series of reactions. He hit upon a system of adding poisons, such as arsenic and cyanide, to interfere with the glycolysis pathway so that there would be a build-up of intermediate metabolites that could then be chemically identified. He also began to isolate the enzymes responsible for each transformation, beginning with one he named *hexokinase*.

Meyerhof was a fabulous interpreter of experimental data, with an unusual ability to paint them into a larger picture. He cut a reserved, rather austere, figure, always dressed in a waistcoat with a gold pocket watch in the left-hand pocket. Hermann Blaschko, another of his young acolytes, remembered him first and foremost as an experimenter:

[Meyerhof] was forced, by the fertility of his own work and that of his colleagues as well as by his own theoretical interests, to spend a good deal of his time at his desk, but one got the impression that he was always impatient to get back to the bench.[46]

Blaschko, like Nachmansohn, was a young Jewish physician with a fascination for biochemistry hired by Meyerhof, initially with no salary or grant money. In the spring of 1926, Meyerhof sent him to London to work in Hill's laboratory for a few weeks. Hill invited him back for one year in 1929 and arranged for him to spend the summer in Plymouth, to meet as many British physiologists as possible.

Karl Lohmann spent most of his time in Meyerhof's laboratory studying various phosphate compounds isolated from muscle.[47] He devised a new technique of heating samples in hydrochloric acid at 100° C for about the time it takes to hard-boil an egg. He then added molybdenum solution to the flask: if the contents turned blue, it indicated that inorganic phosphate had been released from the muscle. At a departmental seminar just before Christmas, he announced that, with a slight modification in technique, he had found the muscle contained phosphoric acid linked to pentose sugar plus adenine, a purine base. He called the new compound *adenylpyrophosphate*, now universally known as ATP (*adenosine triphosphate*). Meyerhof persuaded Lohmann to publish his findings, which he did under the underwhelming title: "On the Pyrophosphate Fraction in Muscle." This has subsequently become one of the most cited papers in the history of biology,[48] and ATP deserves to be as well known as the other molecular triplet fundamental to life: DNA.

In 1929 the International Congress of Physiology was held in the United States for the first time. Boston and specifically Harvard University were the designated hosts. Ernest Starling suggested that a ship should be hired to carry European physiologists across the Atlantic. After Starling died in 1927, the international committee co-opted Hill and placed him in charge of transportation arrangements. He showed all his military organizational zeal in chartering hundreds of places at a discount on the SS *Minnekahda*—a converted, American World War I troopship.[49] On 9 August, about 400 physiologists and their families boarded in the Thames estuary and several hundred more joined later at Boulogne. The slim, white-haired, sun-tanned Hill was at the center of deck games, dinners, and other social activities, always wearing a three-piece suit (Figure 7.6).

Hill delighted in having old friends such as Fletcher, Haldane, and Barcroft as well as members of the UCL department on board with him. Margaret and the children stayed behind at Three Corners. Before sailing, AV wrote to her to say: "I was never so sorry to leave anywhere as I was to leave this morning." On arrival in Boston, he

Figure 7.6 Hill surrounded by fellow physiologists on the SS Minnekahda sailing to Boston for the International Congress, 1929

Photograph courtesy of the Wellcome Collection, London.

Figure 7.7 Albrecht Bethe (1872–1954) using Jakub Parnas (1884–1949) as a pillow on deck
Photograph courtesy of the Wellcome Collection, London

soon met up with Meyerhof and others from Berlin who had sailed with fifty more scientists from Germany and Switzerland. The social activities continued, and at the banquet Hill replied to the speech of Walter Cannon of Harvard—reflecting on the extraordinary size and success of the Congress, which he took as a sign that the center of gravity of modern science was moving away from the shores of Europe.

The large Congress included talks reviewing the broad advances made in recent years, as well as demonstrations in the Harvard laboratories and short ten-minute papers. Lohmann gave one of these to announce his isolation of the novel molecule ATP from muscle. In the audience for his presentation was Cyrus Fiske, the Harvard biochemist who had correctly identified phosphocreatine with his Indian research assistant, Yelapragada SubbaRow. They had continued to look for phosphate-containing compounds in muscle; Fiske, whose self-regard and jealousy were legendary, was angry because he had previously disclosed some of his results to Meyerhof. He suspected that Meyerhof had fed this information to Lohmann, without attribution. He challenged Lohmann after his talk and then insisted that SubbaRow be allowed to present their findings on ATP on the last afternoon of the Congress.[50]

There had been no unseemly rush to publish Lohmann's paper: it appeared in *Naturwissenschaften* only two weeks before the Congress started, whereas he had first announced his discovery to Meyerhof's department at the end of 1928. Priority is a powerful motive for scientists, but there is no reason to believe that the Harvard researchers, specifically SubbaRow, did not make the same discovery independently at around the same time. Given the supreme importance of ATP as the energy currency in biology, many have questioned why Lohmann, who went on to elucidate its structure and many of its properties in Meyerhof's lab, did not receive a Nobel Prize for Chemistry. The explanation may be that Meyerhof was so closely involved that it would have meant him sharing a second Nobel Prize.[51]

Otto Warburg remained in overall charge of the Kaiser Wilhelm Society (KWG) biology institute in Berlin-Dahlem, where Meyerhof had his borrowed rooms. A solitary man with a fixed daily schedule that started with a horse ride early every morning, Warburg had served in a crack cavalry regiment during the 1914–1918 war and was awarded the Iron Cross after being wounded. He retained an imperious manner that made him more at home with military officers than with his fellow scientists. Yet Warburg was driven by his own scientific curiosity, and his output of work was prodigious and varied.[52] He and Meyerhof were only one year apart in age, and both were now functioning as specialist biochemists rather than general physiologists. Despite his superior standing within the KWG, there seems little doubt that Warburg was jealous of Meyerhof's status as a Nobel laureate. It seems that Meyerhof sensed this. In 1928, Warburg's sister recorded in her diary[53] a tense conversation between the pair that she overheard on a staircase at the KWG institute:

MEYERHOF: "What do you have against me?"
WARBURG: "I have nothing at all against your science, but I prefer to choose other people for my personal dealings."
MEYERHOF: "Why?"
WARBURG: "You go to your destination with cold calculations, and I don't like people like that for my personal dealings."
MEYERHOF: "Well, hopefully after this discussion, things will get better between us."

What Warburg did not know was that from 1924, the first year he was qualified, Meyerhof proposed him for a Nobel Prize,[54] and remained his most loyal supporter until Warburg was finally honored in 1931 "for his discovery of the nature and mode of action of the respiratory enzyme."

Notes

1. Alfred Nobel's health and his interest in medicine. NobelPrize.org. Nobel Media AB 2020, www.nobelprize.org/alfred-nobel/alfred-nobels-health-and-his-interest-in-medicine (accessed 25/1/20).
2. M. Teich (1992), 20–21.
3. D. Needham, *Machina Carnis: The Biochemistry of Muscular Contraction in Its Historical Development* (Cambridge: Cambridge University Press, 1971), 41.
4. E. Norby, *Nobel Prizes and Life Sciences* (London: World Scientific, 2010), 28.
5. www.nobelprize.org/nomination/redirector/?redir=archive/ (accessed 27/1/20).

6. Teich (1992), 217–219.
7. Norby (2010), 157–163.
8. Norby (2010), 157–163.
9. Professor Siqvist was more persuasive in his recommendation to the committee that the 1923 prize should be shared equally between Macleod and Banting for their work on insulin—although that proved more controversial over time than the 1922 prize.
10. O.F. Meyerhof to A.V. Hill (29/10/23) AVHL I 3/58.
11. W. Meyerhof (2002), 16–17.
12. Award ceremony speech. NobelPrize.org. Nobel Media, www.nobelprize.org/prizes/medicine/1922/ceremony-speech (accessed 31/01/20).
13. R.F. Foster, *W.B. Yeats: A Life*, Vol. II (Oxford: Oxford University Press, 2003), 245.
14. Angier, I (1978):183.
15. Archibald V. Hill—Banquet speech. NobelPrize.org. Nobel Media, www.nobelprize.org/prizes/medicine/1922/hill/speech (accessed 1/2/2020).
16. Otto Meyerhof—Banquet speech. NobelPrize.org. Nobel Media, www.nobelprize.org/prizes/medicine/1922/meyerhof/speech (accessed 1/2/2020).
17. Archibald V. Hill—Nobel Lecture. NobelPrize.org. Nobel Media, www.nobelprize.org/prizes/medicine/1922/hill/lecture. (accessed 31/1/2020).
18. A.V. Hill, "The Maximum Work and Mechanical Efficiency of Human Muscles, and Their Economical Speed," *Journal of Physiology* 56 (1922): 19–41.
19. Fenn (1923).
20. In a letter to W.H. Bragg (2/6/26), Hill did express a shift in his thinking as he was preparing the Croonian Lecture:

    "I have ceased to be happy about the simple hypothesis of an elastic network containing a viscous material, owing to the extraordinary differences which exist between the apparent viscosities of muscles when stimulated which were previously, when at rest, of just about the same viscosity." WHB archive Royal Institution 3C/45.

21. A.V. Hill, *First and Last Experiments in Muscle Mechanics* (Cambridge: Cambridge University Press, 1970), 2.
22. Otto Meyerhof—Nobel Lecture. NobelPrize.org. Nobel Media, www.nobelprize.org/prizes/medicine/1922/meyerhof/lecture (accessed 13/02/20).
23. Rall (2014), 5.
24. U. Charpa and U. Deichmann (eds.), *Jews and Sciences in German Contexts* (Tubingen: Mohr Siebeck, 2007), 117. Despite the preeminence of German biochemists, many of whom were Jewish prior to 1933, chairs in biochemistry were created very slowly in the universities.
25. O. Meyerhof to O. Warburg (25/4/23) in Werner (1996), 318–321.
26. Peters (1954).
27. Mulley (2009), 311–313, and Angier I:192.
28. Bassett (2002).
29. A.V. Hill, "The Scientific Study of Athletics," *Scientific American* (April 1926): 224–225
30. Ralph W. Gerard (1900–1974) first entered the University of Chicago in 1914, and by 1925 held a PhD and MD. He spent the first year of a research fellowship with Hill in London and the second with Meyerhof in Berlin.
31. R.W. Gerard, "The Minute Experiment and the Large Picture," *Perspectives in Biology and Medicine* 23, no. 4 (1980): 527–540.

32. A.C. Downing, R.W. Gerard, A.V. Hill, "The Heat Production of Nerve," *Proceedings of the Royal Society of London (B)* 100 (1926): 223–251.

33. P. Eggleton and G.P. Eggleton, "The Inorganic Phosphate and a Labile Form of Organic Phosphate in the Gastrocnemius of the Frog," *Biochemical Journal* 21 (1927): 190–195.

34. Needham (1971), 81, where it is stated as about 12,500 cal/g mole when the phosphate group is hydrolyzed.

35. D. Nachmansohn, "Biochemistry as Part of My Life," *Annual Review of Biochemistry* 41 (1972): 1–30.

36. Katz (1978).

37. Angier I (1978):224.

38. Angier I (1978):240.

39. A.V. Hill, *Living Machinery* (New York: Harcourt, Brace & Co), 1927.

40. A.W.H. Bates, *Anti-vivisection and the Profession of Medicine in Britain* (London: Palgrave-Macmillan, 2017), 133–159.

41. A.V. Hill letter to *The Times* (25/11/26) .

42. A.A. Milne, "Vivisection and the Expert" AVHL II 4/61.

43. A.V. Hill, "Three Corners" M&R: 254–258 AVHL I 5/4.

44. A.V. Hill, "The Donnan-Hill Effect," in *The Ethical Dilemma of Science* (New York: Rockefeller Institute Press, 1960), 148–155.

45. Needham (1971), 61–70.

46. Quoted in Peters (1954).

47. Needham (1971), 84–85.

48. M. Emmerich, "Cells Flexing Their Muscles," www.mpg.de/7023399/S005_Flashback_086-087.pdf (accessed 9/3/20).

49. Y. Zotterman, "The Minnekahda Voyage," in W.O. Fenn (ed.), *History of the International Congresses of Physiological Sciences 1889–1968* (n.p.: American Physiological Society, 1968).

50. J.A. Rall, "The XIIIth International Physiology Congress in Boston in 1929: American Physiology Comes of Age," *Advances in Physiology Education* 40 (2016): 5–16.

51. P. Langen and F. Hucho, "Karl Lohmann and the Discovery of ATP," *Angewandte Chemie International Edition* 47 (2008): 1824–1827.

52. Krebs (1972).

53. Werner (1996), 85.

54. It is an interesting reflection on the breadth of Warburg's contributions that Meyerhof first nominated him for his research on photosynthesis and the mechanisms of cell respiration. He cited cell respiration research in all subsequent years, adding his work on tumor cell metabolism at different times. Hill made a small but vital contribution to Warburg's research when visiting Dahlem in 1926. Warburg was studying the effect of carbon monoxide (CO) poisoning on the respiration in yeast cells. He knew that CO would bind to iron containing proteins (most notably hemoglobin) and showed that it had an inhibitory effect on the respiratory enzyme in yeast cells. During a dinner conversation, Hill told him that Cambridge physiologists had shown that the CO binding to the iron-containing heme moiety in hemoglobin was reversed by exposure to light. Warburg left the dinner immediately and went to his lab. He was soon showed that light also protected the yeast cell respiratory enzyme (Atmungsferment) and in a series of beautiful experiments, he demonstrated the absorption spectrum of the enzyme (Krebs, 1972, Gerard, 1980, and Apple, 2021).

8
Revolutions in Physiology and Power

Both Hill and Meyerhof, attempting to picture the energy exchanges that occurred during muscle activity, reached similar analogies. In his Nobel Lecture, Hill stated: "A muscle is like an accumulator, which can be discharged without any kind of combustion or any kind of provision of energy from without: it requires external energy only when it is being recharged." Writing two years later, Meyerhof compared carbohydrate metabolism to the operation of a storage battery, saying that anerobic glycolysis was akin to discharging it while oxidative resynthesis was like recharging it.[1] As mentioned before, Meyerhof was deeply attached to the lactic acid theory of muscle contraction, where not just the fuel necessary for muscle activity but also the change in muscle fiber length was due to lactic acid. Just as with Hill's pet visco-elastic theory, Meyerhof's favored ideas were upended by new experimental findings.

The demise of the lactic acid theory brooked no argument following a series of papers, beginning in 1930, from Einar Lundsgaard—a tall, pipe-smoking physiologist from the University of Copenhagen. His primary interest was in the biological properties of specific amino acids. Chemical analysis of thyroid hormone (thyroxine) had just shown that it contained iodine, and Lundsgaard decided to investigate the metabolic effects of iodoacetic acid. He neutralized the acid before giving intravenous injections into rabbits. The animals would behave normally for a short period before falling on their sides, their breathing would stop, and they became as stiff as a board.[2] He found that if he cut the nerve to one limb prior to the injection, the rigidity spread through the rest of the body but that limb was spared. However, if the cut nerve to that leg was then electrically stimulated, after a short series of contractions, the limb would also stiffen. The rigor (stiffness) was brought on by muscle activity itself, yet Lundsgaard found there was no formation of lactic acid in the poisoned muscle (nor in the denervated limb after electrical stimulation). In his first paper, "Researches on Muscle Contraction without Lactic Acid Formation," Lundsgaard stated: "This result can scarcely be explained in any other way than that the poisoned muscles have performed work without lactic acid formation."[3] Lundsgaard "extended [his] experiments to the different phosphorus fractions of the muscle" and analyzed the breakdown of phosphocreatine. He found all the phosphocreatine in poisoned muscles was used up during the contractions, whereas, after activity, normal muscle showed a decrease of only about one-quarter in its phosphocreatine content. This led Lundsgaard to conclude that lactic acid formation normally leads to a resynthesis of phosphocreatine, while the breakdown of phosphocreatine releases energy for contraction.

Bound by Muscle. Andrew Brown, Oxford University Press. © Oxford University Press 2023.
DOI: 10.1093/oso/9780197582633.003.0008

Meyerhof had moved from Berlin-Dahlem at the end of 1929 to a purpose-built Kaiser Wilhelm Society institute for medical research in Heidelberg. His old chief, Ludolf Krehl, was in overall charge of the institute, as well as being head of the medical department. Having failed to recruit Otto Warburg, Krehl took his advice and appointed Meyerhof as head of the physiology department. Meyerhof was able to influence the design of its laboratories, set in wooded parkland near the Neckar River. He would cycle from his new house, also built by the Kaiser Wilhelm Society (KWG), to the institute. In accordance with Meyerhof's characteristic impatience for data to be generated, his research team was expected to stand and walk around on builders' planks as soon as water and gas were piped into the laboratories before the floors were laid.

Lundsgaard wrote to Meyerhof to inform him of his findings about muscle contraction without lactic acid production before they were published. Meyerhof, sensing how important Lundsgaard's findings were, invited him to Heidelberg to continue his research. It is to the credit of both men that they agreed to this collaboration since Meyerhof, through his published work, was the leading proponent for the widely accepted lactic acid theory. A smaller-minded scientist than Meyerhof might have worried about loss of reputation: a less confident young man than Lundsgaard might have ducked the opportunity to demonstrate his findings in the lions' den, although replication there would end any doubt. Hermann Blaschko described the moment that Lundsgaard's letter was read out:

> [Meyerhof] told of Lundsgaard's letter describing his still unpublished experiments, according to which lactic acid production was not essential to muscular contraction. He was expecting Lundsgaard in the laboratory in the coming month, but although reserving final judgment until Lundsgaard's arrival, he was quite prepared to believe that Lundsgaard was right. I, who earlier on had witnessed Meyerhof's critical attitude to Embden's experiments on the time course of lactic acid formation, still remember his surprise at the readiness to consider these new and—to Meyerhof—still unverified observations.[4]

Fritz Lipmann, another member of the research group who had moved from Dahlem to Heidelberg, recalled that Lundsgaard's "startling news ... [was] very upsetting to our group which looked upon glycolytic lactic acid as the link between metabolic energy generation and muscle contraction."[5]

Lipmann met Lundsgaard at the station and took him and Mrs. Lundsgaard straight to the physiology department from the train. Over the next six months, Lundsgaard extended his own findings, while several other lab members (Blaschko, Lipmann, Lohmann), often in conjunction with Meyerhof, were drawn into related experiments. Lundsgaard found that poisoned muscles working in the absence of oxygen deplete their phosphocreatine content more than when oxygen is present.[6] He concluded, therefore, that at least in iodoacetate poisoned muscle, there is "oxidative resynthesis" of phosphocreatine. Lundsgaard believed that the hydrolysis (chemical breakdown) of phosphocreatine yielded the energy for muscle contraction (it was "nearer" to the conversion of metabolic energy into mechanical energy than lactic acid was), but he thought there could still be "a third unknown reaction which may be directly coupled with the course of the contraction."[7]

Naturally, British physiologists were interested in the novel findings coming from Heidelberg, and there was some collaboration between Hill's group at UCL and William Hartree in Cambridge to investigate the heat production during the stimulation of poisoned muscle. Hill obviously communicated their results to Lundsgaard, back in Copenhagen, and he replied early in 1931:[8]

> It is indeed very strange that there is no difference at all to be found between normal and poisoned muscles in respect of initial heat. I should say that this circumstance demonstrates that the lactic acid formation only plays quite a secondary role for the liberation of energy *during* the contraction.

Hill visited Meyerhof's new department during 1931 (Figure 8.1) and subsequently gave an address on "The Revolution in Muscle Physiology" to the BAAS.[9] He dated the origin of the revolution to the last day of 1926, when the Eggletons sent

Figure 8.1 Hill's visit to Meyerhof's new department in Heidelberg, 1931. Standing behind Meyerhof and Hill (left to right): K. Lohmann, A. von Muralt, G. Benetato, H. Blaschko, A. Grollman, H. Laser, Miss Wagner, W. Schultz (lab tech), and E. Boyland Photograph courtesy of the Archives of the Max Planck Society, Berlin.

their paper on phosphagen [phosphocreatine] for publication. He was convinced that subsequent findings had changed the accepted view of the relationship between the chemistry and the energy exchanges involved in muscle activity, but "[N]o theory of muscle contraction had been invented." While two German physiologists had suggested Hill's ideas about isolated muscle physiology had been "completely shattered by the recent research," he reassured his audience that previous experimental results were still valid and explained some of his previous findings in terms of the breakdown and resynthesis of phosphocreatine.

In 1932, Meyerhof and Lohmann measured the heat released by the hydrolysis of one phosphate unit from ATP, converting it to adenosine diphosphate, ADP. They found a tremendous quantity, about 25,000 calories per gram-molecule, so that ATP was a compound unequaled in terms of its stored energy. Indeed, it showed about double the heat of hydrolysis of phosphocreatine, leading them to suggest that the chemical energy released from one molecule of ATP→ ADP might permit the synthesis of two molecules of phosphocreatine.[10] Was this the third unknown reaction that Lundsgaard had wondered about? In 1934, Lohmann reported that while there was no enzyme present in muscle that could remove the phosphate group from phosphocreatine, there was an enzyme that he named *creatine kinase* that could transfer a phosphate group to ADP, forming ATP. The reaction (which is reversible depending on the metabolic state of vertebrate muscle) has become known as the Lohmann reaction:

$$\text{Phosphocreatine} + \text{ADP} \leftrightarrow \text{Creatine} + \text{ATP}$$
$$\uparrow$$
$$\textit{Creatine kinase}$$

Lohmann suggested that if his results obtained with dialyzed muscle extract applied to intact tissue, there would be a series of chemical reactions during muscle contraction: (1) the breakdown of ATP; which (2) is reversed by the breakdown of phosphocreatine. The resynthesis of phosphocreatine follows; (3) partly anaerobically through the energy of lactic acid formation, partly aerobically through oxidative processes (usually carbohydrate combustion); these oxidative processes bring about the resynthesis of lactic acid to glycogen in the Pasteur-Meyerhof reaction.[11] In 1935, Lohmann established the chemical formula of ATP that is still accepted today. The ATP molecule is the universal energy currency in cells and needs to be available instantly in muscle, for example, when there is a sudden increase in activity.[12] It is not stored in substantial amounts: when there is increased demand, the reaction (above) catalyzed by creatine kinase ensures that ATP is efficiently and instantly replenished from the phosphocreatine that is stored in muscle cells.

During the 1930s many biochemists around the world, notably Gustav Embden, Jakub Parnas, and Otto Warburg, made important contributions to the detailed uncovering of the ten enzymatically powered steps of glycolysis that convert glucose to pyruvate, with the creation of two molecules of ATP. In muscle there is an eleventh enzyme, *lactate dehydrogenase*, that converts pyruvate to lactate.[13] The overall process of anaerobic glycolysis in muscle can be summarized:

$$Glucose + 2ADP + 2Phosphate \rightarrow 2\ Lactate + 2ATP + 2H_2O + 2H^+$$

The investigation of compounds with high-energy phosphate bonds present in muscle led to a clearer understanding of the energy exchanges involved in glycolysis. It is a mark of how much progress had been made, when one remembers that in 1912, Hill made the apparently reasonable suggestion that there was just one chemical step from a precursor molecule to lactic acid, based on the high speed of insects' wing beats. Although other laboratories were active participants, it is generally acknowledged that the "lion's share" of the individual enzymes and intermediary substances responsible were first identified in Meyerhof's department: glycolysis is often referred to as the Embden-Meyerhof pathway.[14]

From the time of his inaugural lecture at Kiel in 1911, Meyerhof had consistently promoted the idea that there was a unity in the chemistry supporting different life forms. In the mid-1930s, he was able to confirm this theory in detail by studying glycolysis in bacteria, yeast, as well as muscle from a variety of species. He found the theory to hold with minor exceptions: for example, he and Lohmann showed that some invertebrate muscles, such as those in lobster, contain phosphoarginine instead of phosphocreatine. Meyerhof believed that this distinction resulted from an evolutionary chemical mutation, but more recent studies suggest that this is an oversimplification.[15]

Meyerhof spent most of his waking hours in the laboratory, inspiring his juniors to follow him on the labyrinthine path to the frontiers of biochemistry. His department was divided into two floors, with the visiting research workers in a large room equipped with chemical benches on the first floor; on the second floor were the private laboratories and offices for Meyerhof and Lohmann on the second floor. On entering the laboratory, Professor Meyerhof would make the rounds of the communal lab, asking each worker in turn the same two questions every day: "What did you do yesterday?" and "What will you do today?"[16] Alexander von Muralt, a Swiss physiologist who spent five years in Meyerhof's laboratory, recalled late morning conversations with the "chief": "lively remarks and answers, his vigorous agreement or his cool disapproval, his facial expression alternating between roguishness and seriousness and reflecting the nuances of his rapid thinking, were an unforgettable experience in the exchange of ideas with a great man."[17] His rounds complete, Meyerhof would ascend to the second floor, where he did not welcome interruptions or questions from the researchers downstairs. He would spend half the day conducting his own experiments, often with the assistance of his long-serving and trusted laboratory technician, Walter Schultz. The remainder of the day would be spent writing papers. Lohmann remembered a cordial relationship with Meyerhof, who was very silent by nature.

Individual physiologists from overseas would still come to spend short period in Hill's laboratory, but he never built a team of talent such as the one assembled in

Heidelberg. Instead he preferred to rely on his loyal staff, A.C. Downing, the instrument maker, and J.L. Parkinson, his laboratory manager, both of whom were supported by the Foulerton Fund of the Royal Society.

One regular American visitor who became a great friend was Detlev Bronk, recently appointed to a new chair in biophysics at the University of Pennsylvania. In 1930, Bronk invited Hill to Philadelphia to receive an honorary doctorate and to give a series of lectures, subsequently published as *Adventures in Biophysics*. In June, Hill received a letter written in impeccable English from a Chinese scientist, T.P. Feng, who was studying nerve hypoxia in Chicago with Ralph Gerard (who assisted Hill in the first successful measurement of the heat generated by a nerve impulse). Feng had recently been instructed by his professor in Peking to pursue a PhD at UCL instead of Chicago. The professor also contacted Hill, who wrote back to Feng saying, "If you as good as Lim says you are, come along."[18] Feng arrived in London that fall and spent the next three years studying heat production in muscle and nerve, including one summer at the MBS in Plymouth. He became a favorite of the Hill family, and AV made sure he spent time with England's leading neurophysiologists—Sherrington in Oxford and Adrian in Cambridge—before arranging for him to return to the United States in order to spend time with Bronk in Philadelphia. It was typical of Hill to go out of his way to support a young scientist whom he regarded as talented; just as we have seen, he played a major role in finding unexpected opportunities for ex-Brigands such as E.A. Milne and William Hartree.

Hill was becoming a man of public affairs by accretion. He had been a governor of Blundell's School since the end of the war and added similar positions at Harrow and Highgate schools in 1929. In 1931 he was elected to the Council of the Royal Society, two years later becoming its Biological Secretary. He was also a committee member of the International Congress of Physiology and centrally involved in the 1932 gathering in Rome. AV traveled there by train with seventeen-year-old David, who about to enter his last year at Highgate school. The official invitation to the Congress reflected the pervasive influence of Benito Mussolini, the vainglorious leader of the ruling Fascist Party:

> His Excellency, the Head of the Italian Government, has instructed Professor Filippo Bottazzi to organize the XIVth International Congress and has decided that it should take place in Rome from August 29th to September 3rd].

Il Duce's Fascist Party took its name from *fasces*, the ancient Roman emblem of a bundle of sticks with a protruding axe—one was depicted on the reverse of the conference medal. The trappings of a one-party state were obvious to the nearly 1,000 delegates, who made their way to the Campidoglio—they passed several cordons of police and had their credentials examined by multiple times before entering the Sala di Giulio Cesare. Two junior colleagues from UCL wrote that the "the circumstances [were] embarrassing to the law-abiding scientific worker."[19] Hill was sharing the stage with Il Duce, who shook his hand warmly before Hill began his inaugural address. He earned the gratitude of the audience by not delivering his preprinted lecture on the production of heat in nerves and muscles. With a "bluntness tempered by his usual good humour," he decried what he saw as the unnecessary proliferation of scientific literature, and, more curiously, the wasteful nature of scientific meetings. He wrote to Margaret that Mussolini "again shook my hand warmly and said how much he had appreciated my lecture, which he had understood."[20] Despite the authoritarian political atmosphere in Italy, it was a pleasurable conference, with interesting social visits (including an audience with the Pope for AV as part of a small group), but as if to reinforce the words of their senior colleague, Bayliss and Eggleton concluded there was an "absence of any advance of a notable kind in any quarter." Hill's Nobel co-laureate, Meyerhof, shied away from the limelight and contented himself with taking his juniors to "a delightful meal at Alfredo's."[21]

The Hill family was at its most cohesive in the summers spent at Three Corners. AV would take his children to the MBS in Plymouth, where they would be able to go out on boats with the station staff. Their friends and cousins often visited, as did AV's scientific colleagues and their families. Gottfried and Bettina Meyerhof both came to stay for one month in the summer of 1933. Margaret's favorite guest was Norah Clegg, her Manchester friend, who moved to London in 1933. Eglantyne Jebb died in Geneva in 1928, and Norah would become Margaret's confidante and traveling companion throughout the 1930s and beyond. When in London, Margaret was involved in local authority politics as a member of Hornsey Council. She also founded the Hornsey Housing Trust, which acquired old houses and converted them into low-rent apartments for the elderly.

The Hill children displayed the vicissitudes of teenage life. Polly spent time in Germany to improve her language skills before going up to Cambridge; there she read geography for just one term before switching to economics. David, who loved running and target shooting, now followed in AV's academic footsteps by obtaining a major scholarship to Trinity College Cambridge to study natural sciences, taking physiology for the Part II Tripos. Janet was a spirited girl who took no prisoners, while Maurice's adolescence was marked by indolence and illness. AV made what was becoming his annual voyage to the United States in June 1933 for a meeting of the American Association for the Advancement of Science (AAAS) in Chicago. Before leaving, he paid a visit to his friend and former tutor, Sir Walter Fletcher, who was bedridden at his house in Chelsea. When Hill arrived in New York, he was told of Fletcher's death—a deeply felt loss.

While AV was not disturbed to be in Mussolini's company in Rome, Hitler's assumption of power in early 1933, with the subsequent violent repression of all political opposition and unchecked anti-Semitism, filled him with foreboding. The burning of the Reichstag building at the end of February gave the Nazis carte blanche through the enactment of the Enabling Law to suppress and victimize whomever they chose. In early April, a civil service law was passed that applied to all German universities because they were (and are) public institutions. It called for the dismissal of officials, including most university academics, if they had at least one Jewish grandparent or were political opponents of the regime.[22] While this was reported to a certain extent in the British press, the impact was greater the closer one was to Germany, where the thuggish Brownshirts were already beating Jews in the streets.

William Beveridge, the social scientist and director of the London School of Economics, was at a conference in Vienna in April. He was joined for tea in his hotel by the peripatetic and indefatigable Hungarian physicist, Leo Szilard. Szilard was already planning how Jewish intellectuals, who were being dismissed in large numbers from German universities, might be supported in England and other countries. He pressed Beveridge to help and a few days later popped up in London, where he continued to badger Beveridge by letter and through visits to the LSE.[23] He also visited UCL, where he found Hill to be a ready listener. Indeed, Szilard told him he was contemplating leaving physics for biology, and AV offered him a part-time position as a demonstrator so that "by teaching physiology, you would learn physiology."[24]

Beveridge visited Cambridge in early May and suggested to Rutherford that British scientists and intellectuals should do something to relieve their German colleagues. A few days later Rutherford, Hopkins the biochemist (who had succeeded Rutherford as president of the Royal Society), and Hill met at Trinity College.[25] They decided the matter would be considered at the next Royal Society Council meeting on 11 May. Together with Sir Henry Dale, Hill's predecessor as the biological secretary, the Council approved an appeal to organize assistance for scholars who had been forced to relinquish their posts in German universities. The statement, signed by forty-one prominent British academics including John Maynard Keynes and Hill, proclaimed the formation of the Academic Assistance Council (AAC) to seek funds to assist university teachers who "on the grounds of religion, political opinion, or race, are unable to carry on their work in their own country."[26] Beveridge persuaded Lord Rutherford to become the president of the AAC; Hill was elected to be one of two vice presidents at its first meeting at the end of May. The Royal Society provided office accommodation in Burlington House, and UCL was well represented from the outset. Although the persecution of Jewish scholars in Germany was the most pressing issue, the AAC claimed not to be unfriendly to the people of any country and not to be sitting in judgment on other forms of government. "Our only aims," the group stated, "are the relief of suffering and the defence of learning and science." Hill's colleague at UCL, Charles Singer, a humanist and historian of science, had been responsible for drafting some of the early AAC statements, but the Royal Society advised that it would be impolitic for him as a Jew to be appointed as general secretary, although he remained a member of the executive committee.[27] By the summer, the AAC had raised almost £10,000, a quarter of which comprised a block grant from the Central Jewish Fund. By the end of the year nearly all the money was allocated, mostly in the form of annual grants of £250 to married scholars.[28]

Walter Adams,[29] a history lecturer at UCL, agreed to serve as general secretary for the AAC. His assistant secretary, who would become the linchpin of the organization, was Esther "Tess" Simpson (Figure 8.2). Originally from Leeds, she was working for the International Student Service in Vienna, where she mixed in musical

Figure 8.2 Esther "Tess" Simpson (1903–1996) Humanitarian, organizer, violinist
Photograph by Lotte Meitner-Graf, London. Copyright © The Lotte Meitner-Graf Archive—info@
lottemeitnergraf.com (Original print courtesy Leeds University Library).

and intellectual circles. She met Szilard during his fleeting visit to the city and oblig-
ingly typed several letters for him. A few weeks later, she moved to Geneva, where
she encountered Szilard again. He offered her a job with the AAC, "with the chance
to help the sort of people I'd played chamber music with in Vienna."[30] Despite taking
a two-thirds cut in salary, she accepted immediately and started work in the two attic
rooms provided by the Royal Society in July.

In mid-September, Adams drove up to the north Norfolk coast in search of the
Jewish physicist most reviled by the Nazis, the outspoken critic the regime most
wanted to eliminate. Albert Einstein, who renounced his German citizenship and
resigned from the nation's most prestigious science academies soon after Hitler came
to power, had taken unlikely temporary refuge in a small holiday hut owned by a
Conservative MP, Commander Oliver Locker-Lampson.[31] Having passed two young
women armed with shotguns, Adams found the windswept professor walking around
a "little hedged compound." He asked Einstein if he would speak on behalf of German
academic refugees at a meeting that the AAC was helping to organize. Einstein con-
sented and Locker-Lampson rushed to the nearest telephone to book the Albert Hall.[32]

The meeting on "Science and Civilization" took place on 3 October with Einstein as the star speaker.[33] There was an enthusiastic audience of 10,000, with about 1,000 London university students in their gowns acting as stewards. The *Daily Mail*, in an editorial, predicted that although the meeting was intended to raise funds on behalf of German refugees, it would be seen as a demonstration against Hitler and the Nazi regime. Lord Rutherford gave a measured opening address, explicitly downplaying any political antagonism and appealing for generous donations, "to save these academic professional workers who are now dependent on the charity of humanity." Einstein also refused "to act as scourge of the conduct of a nation which for many years has considered me as their own" but warned that without intellectual and individual freedom: "There would be no comfortable houses for the mass of the people, no railway, no wireless, no protection against epidemics, no cheap books, no culture and no enjoyment of art for all." The meeting raised a few thousand pounds, but Locker-Lampson insisted on retaining an undeclared amount for future activities, "an irregularity," as Beveridge wrote to Adams, that "has caused a great deal of irritation and distress."[34]

Hill found his own more strident voice on the Nazi threat to intellectual life, when delivering the Huxley Memorial Lecture at a meeting of the BAAS in Birmingham on 16 November.[35] He referred to the permissive attitude and tolerance shown generally by civilized societies toward those concerned with scientific discovery and learning. It was the universality of the scientific method which yields reproducible results—independent of custom, opinion, emotion, race, sex, or nationality—that Hill believed gives science a unique place among human endeavors. To preserve their privileged place in society, scientists should not overstep their areas of expertise on broader political or moral issues. They were, however, entitled to express their views on such matters as long as they did not expect or demand that they were taken more seriously than "more ordinary people." Equally, science should not be used for propaganda purposes: "The coercion of scientific people to certain specified political opinions, as in Russia, Germany, or Italy, may lower the standard of scientific honesty and bring science itself into contempt." Broadening his argument to note the waning of religious and political toleration, Hill regretted the recent reversal of "progress" so that "gentleness has ceased to be admired: communism, and its natural—its inevitable—antibody, fascism, have taken charge of the minds of a large section of human society." He was stunned and saddened by the rapidity of Germany's fall from grace. Whereas he had believed that the intellectual cooperation of Germany was a necessity in setting science on an international basis, he was now thoroughly disillusioned.

The facts are not in dispute. I speak with some knowledge, having a personal acquaintance with, and having recently seen, many of the victims of the Aryan Myth. Apart from thousands of professional men, lawyers, doctors, teachers, who have been prevented from following their profession; apart from tens of thousands of tradesmen and workers whose means of livelihood have been removed, apart from 100,000 in concentration camps, often for no cause beyond independence of thought or speech, something over a thousand scholars and scientists have been dismissed, among them some of the most eminent in Germany. ... It is difficult to believe in progress, at least in decency and commonsense, when this can happen almost in a night in a previously civilized state.

In the same month that Hill gave the Huxley Memorial Lecture, Johannes Stark, the 1919 Nobel Laureate in Physics, oversaw a celebration of the "Commitment of the Professors at German Universities and Colleges to Adolf Hitler"[36] in the Alberthalle in Leipzig. Stark had been unsuccessful in obtaining the plum positions in German physics both before and after he struck it lucky in 1919, largely because his mathematical ability was limited, and because he did not embrace the new quantum theory.[37] His failure to ascend to the summit led to a burning resentment of those more successful, especially Einstein and his Jewish colleagues. As early as 1924, with his fellow Nobel Laureate and anti-Semite, Philipp Lenard, Stark composed a paean to the "Hitler spirit and science." The pair would become the leading proponents of "Aryan physics." Stark's reward came in 1933, when he was appointed a president of the Physical and Technical Institute of the German Reich in Berlin after the dismissal of Friedrich Paschen, who had been so generous to Hill on his first visit to Germany in 1911. Stark read the abridged version of Hill's lecture that appeared in *Nature* in late December. As a leading advocate for Aryan science, he could not let it pass unchallenged in such a prestigious, international journal.

Stark's critical rejoinder to the Huxley Memorial Lecture was published in *Nature* two months later.[38] He accused Hill of presenting an inaccurate account of the Nazi government's treatment of scientists and sought to excuse the impression of deliberate anti-Semitism as a necessary attempt to curtail the Jewish near monopoly of academic posts. Stark stated there were fewer than 10,000 in concentration camps, and far from being held over freedom of thought or speech, they were guilty of high treason or other pernicious acts. He admonished Hill for mixing politics with science, while ignoring "the first duty of a scientist, which is conscientiously to ascertain the facts before coming to a conclusion." *Nature's* editor, Sir Richard Gregory, showed Hill Stark's personal attack and made sure his rebuttal appeared on the same page. Dismissing Stark's anti-Semitism as absurd, Hill disputed his assertion that no Jews who served in the war had been removed from their posts. He also took issue with the notion that some Jewish scientists had just chosen to leave, pointing out that they left because they found it impossible to carry out their work in Germany. Hill closed with an appeal for donations to the AAC—"for in spite of all the quibbles, scholars and scientists are still being dismissed."

Stark was not prepared to let the matter rest and wrote to Rutherford, complaining that Hill's letter contained "gross falsehood of a political and cultural kind ... against Germany."[39] Rutherford discussed Stark's letter with Gregory, but not apparently with Hill. Rutherford sent a lengthy reply on 14 March, which began in an emollient tone but ended unambiguously: "We all sincerely hope that this break with the traditions of intellectual freedom in your country is only a passing phase and does not indicate any permanent change of attitude towards the freedom of science and learning." Gregory also wrote to Stark, and informed Hill that Stark had written another letter to *Nature*.[40] In his letter, Stark claimed that the National-Socialist Government has not subjected Jewish scientists to exceptional treatment or forced them to emigrate: "it has passed a law for the reform of the Civil Service which applies to all kind of officials, not only those concerned with science."[41] The perfect riposte came from J.B.S. Haldane, a biologist at UCL:

> Prof Stark's letter in Nature may not prove convincing to all its readers. The fact that non-Aryans have been expelled from other posts does not necessarily justify their expulsion from scientific positions unless the premise that "two blacks make a white" has first been conceded.[42]

Gregory agreed that Hill did not need to respond to Stark directly, and he contented himself with another appeal for funds on behalf of the AAC.

The *Nature* correspondence, together with the summary of the Huxley Memorial Lecture, made a "tremendous impression" on one medical student about to qualify in Leipzig. Bernard Katz was already experiencing Nazi oppression: he was not allowed to collect the physiology prize he had been awarded on "racial grounds" and was made to resign from the students' union.[43] He thought about emigrating to Palestine and, in the summer of 1934, was able to discuss his plan with the Zionist Chaim Weizmann, who was setting up a research institute in Rehovot. Weizmann, a notable professor of chemistry at Manchester University, hearing of Katz's admiration for Hill, wrote a recommendation for him to his former colleague. Katz arrived at UCL to be Hill's PhD student early in 1935 (Figure 8.3). At their first meeting, Katz "was struck by his handsome, somewhat military appearance, very tall and youthful looking, with contrasting grey hair."[44]

Like all visitors to Hill's lab, he was shown a toy figure of Hitler with a movable saluting arm that had been donated by Meyerhof's son, Gottfried, now an engineering student at UCL. Hill explained he was grateful to the Führer for all the excellent people he was sending to England to work. Parkinson, the laboratory manager (Figure 8.4), kept a poster of Hitler on the wall and would hand refugees a toy gun and tell them to shoot.[45] Katz moved into Hurstbourne and became a member of the Hill family for several years.

One of the first physiologists to be rescued from Germany was Hermann Blaschko from Meyerhof's group. He had spent 1929–1930 working as a biochemist at UCL, but in 1933 he had been in hospital in Freiburg with a recrudescence of pulmonary TB.[46] Out of the blue, he received a letter from Hill, who knew he was unwell, reminding him he could always come back to London. He arrived in May 1933, one of the first AAC scholars, and spent the remainder of the year helping with the arrival of others before moving to the Cambridge physiology department, supported by an AAC grant. Blaschko remembered Hill as being "marvellously patient" in helping others find suitable work. Of course, Hill had to cast his net wider than just those he

Figure 8.3 Bernard Katz (1911–2003)
Portrait of Bernard Katz by unknown artist, 1950.
© The Royal Society.

knew personally in Germany, and he used Meyerhof as a guide to suitability of potential émigrés.[47] Although the AAC supported a wide range of scientists, Hill placed particular emphasis on biological scientists, preferring the British university system of first- to third-class rankings. He agreed with Meyerhof that Hans Krebs[48] was first-class, but Hill was also interested in second-class pharmacologists and physiologists, removed from their posts. Naturally, he made occasional errors of judgment, as when he wrote about Ernst Chain that "he is a most objectionable fellow" before admitting that he was not qualified to assess his scientific qualifications.[49] Despite Hill having "no doubt that he is an exceedingly difficult person," Chain was placed in the chemical pathology department at UCL when he first arrived in London, before going to Cambridge to work in Hopkins's department.

Academic ability was the main criterion for deciding which candidates the AAC would support. In July 1933, Hill wrote to the Committee of the Physiological Society saying he had been asked "to get information about a number of German workers in Medicine or the Medical Sciences,"[50] requesting it to add such details as the individuals' present posts, scientific standing, research interests, and where they might find

Figure 8.4   J.L. Parkinson. Long-serving laboratory manager at UCL
Photograph courtesy of the Wellcome Collection, London.

suitable opportunities in this country. The Rockefeller Foundation (RF), which had funded German biology research generously since 1919, also supplied data on the scientists it supported. The AAC's effort was expanding and destined to evolve beyond an ad hoc response to what many hoped would be just a temporary aberration in Germany. Hill displayed a remarkable sensitivity in his understanding of the plight of the dismissed academics. He wrote to Beveridge on New Year's Day 1934:

> It is not that these people will perish as human beings, but that as scholars and scientists they will be heard of no more, since they will have to take up something else in order to live.[51]

At the same time, Hill retained a perspective that went beyond the émigrés as individuals, telling Beveridge what was needed was "an organization to stand for the integrity, independence and the universality of science, with money behind it."[52] He grasped more quickly than many of his colleagues that the advent of fascism in Europe not only threatened individual scholars, it could also undermine the ethos of learning and education vital to civilization. Influenced by Hill, Beveridge was especially concerned that there needed to be a permanent organization to protect scholarship, whenever it came under political attack, from whatever quarter. Reflecting this wider purpose, in April 1936 Lord Rutherford officially announced a change in the title of the AAC to the Society for the Protection of Science and Learning (SPSL). He explained:

> The devastation of German universities still continues; not only teachers of Jewish descent, but many others who are regarded as "politically unreliable" are being prevented from making their contribution to the common cause of scholarship.[53]

Hill with two other SPSL stalwarts wrote an article for *The Times* in which they observed: "A German university under the new regime is as to one part a regiment preparing for war, as to the other an intellectual concentration camp."[54]

In addition to his other responsibilities, Hill remained a member of the international committee that organized the triennial physiology congresses. The fifteenth conference was due to be held in the USSR in the summer of 1935. The chairman of the committee, Professor Johansson from Stockholm, resigned on the grounds of ill-health shortly before, recommending that Hill should replace him. Once again, Hill organized a ship to transport the cream of British physiologists. He took Margaret and their two oldest children along with him. Polly, who was reading economics at Newnham College, Cambridge, kept a diary[55] of the trip that is full of acerbic observations.

They embarked from the pool of London on a dirty Soviet ship, the *Smolny*. The captain was shabbily dressed and had a shaven head; he was not allowed off the ship because of multiple imprisonments in the USSR. When the ship cast off, there was no sign of Professor and Mrs. Barcroft, who were later found asleep in their cabin. Dinner consisted of weak tea and caviar. After crossing the North Sea, the ship sailed up the Kiel Canal, and all along the bank children gave the Hitler salute and shouted, "Heil Hitler!" At Kiel, the distinguished cardiovascular physiologist Otto Frank boarded with his wife. He had been forced to retire from his chair in Munich the previous year. He reminded Polly of "a meek form of Dr. Dolittle" and his wife seemed like a nursemaid. As the ship steamed toward Leningrad, the wireless operator interviewed AV, Barcroft, and the neurophysiologist Edgar Adrian in order to forward their opinions. One question they ducked was—"Who is the greatest physiologist?"—simply acknowledging Pavlov was the most famous. Hill's feeling was that the Congress would have the same impact on the Russian people as the Olympic Games would on the British in their own country. Leningrad was *en fête*, with red banners everywhere in the streets, but that did not manage to hide the dilapidated state of its buildings. The Hills were issued with blue passes that gave them access to the Hermitage Museum, free tram rides, and extra food.

There were at least 1,200 registered delegates to the Congress, 500 from the USSR, often swollen by many hundreds more at social events.[56] Polly realized early on that "The Congress consists of Pavlov, who is already a legendary figure, and everyone else who is not Pavlov." The great man was eighty-six years old, and Polly, like everyone, was charmed by his "modesty and simplicity." The first informal social reception was held in the marble hall of the Ethnographical Museum. Polly noticed her father misbehaving with Banting—"like two naughty schoolboys"—as they dropped paper onto Charles Best's head from the gallery to get his attention. At 10 p.m., the first magnificent banquet of the week was set on one hundred tables, with an orchestra playing until the early hours.

The Hills stayed with Lady Muriel Paget, a remarkable philanthropist and relief worker, who lived in what had been the Swedish Legation. She had been responsible for setting up the Anglo-Russian Hospital in the Dmitri Palace during the 1914–1918 war. After the war, with some assistance from Eglantyne Jebb and Save the Children, she had set up free clothing, food kitchens, and clinics for children in Russia and neighboring countries.[57] She was now caring for displaced British persons, who had been unable or unwilling to leave after the Bolshevik revolution. She had no illusions about the horrors of the new regime and told Polly that 200,000 people with suspected foreign connections had been liquidated in the city since the beginning of the year (subsequently reducing her estimate to 30,000).

The Congress proper began at 11 a.m. the next morning in the Uritsky Palace, with a presidential address from Academician Pavlov, who was flanked on the platform by Hill and Louis Lapicque, as representatives of the international committee (Figure 8.5).

Polly described walking through the hall with her father as like "dragging him through treacle." She listened to a simultaneous English translation via headphones while Pavlov spoke of the Soviet desire for peace. After speeches by various elderly Russian functionaries, Walter Cannon from Harvard delivered an impassioned address. Before starting his lecture on the chemical transmission of nerve impulses, he discussed the damaging effects of politics on science and scientists. He contrasted the funds and social importance accorded to science in the host country with the attitudes of other governments. This was front-page news the next day, with the Soviet journalists seeing it as a reference to the way science was suffering during the economic depressions of the capitalist countries. The newspaper coverage reassured the German physiologists that Cannon was not directly criticizing them so that "the Germans nearly went home, but felt better when they understood it was not a personal offence."[58]

Lady Muriel gave a dinner party for the Hills. When they entered the dining room, Margaret noticed a diminutive lady, who appeared to be wearing a coat made of

Figure 8.5 Ivan P. Pavlov (1849–1936) addressing the opening session of the International Congress of Physiology, Leningrad 1935, with Hill seated to his left
Photograph courtesy of the Wellcome Collection, London.

curtain lace. The woman spoke to her in German and said, "You are a Keynes." She was the mother of Lydia Lopokova, the internationally famous ballerina who married Maynard Keynes. She had been given permission by the secret police to meet the foreign visitors. A few days later Margaret was permitted to visit her in the one-room flat that she shared in grim poverty with Lydia's sister, also a retired ballerina. Hill's book *Adventures in Biophysics* had sold 3,000 copies in the USSR, so there were some spare rubles to spend in the shops. Although bread rationing had finished at the end of 1934, there were still long queues for fruit and vegetables that were weighed out with grudging precision. Polly sensed the economic hardships stemmed from insufficient production rather than poor distribution.

Sunday, 11 August, was a rest day for the Congress and visits were arranged to a wide variety of sites. David and Polly went to several farms, a bread factory and a shoe factory. Polly, an astute economist who would later make her name studying African states, decided there was no valid way to calculate exchange rates and that the Soviet Union could not be compared with any other country, "only with itself through time." AV was taken to athletic facilities and physical training colleges, while

Margaret saw public housing projects and children's clinics. In the evening they were all taken in Lincoln limousines to the summer palace built by Peter the Great. The 30-mile journey took them along roads lined by crowds of people waving at the convoy of immaculate cars. The guests arrived back to Leningrad in time for a lavish dinner, starting at midnight.

Two days later, after a full plenary session of the Congress and an evening of music and ballet at the Academic Theatre, AV and Margaret were guests at an eight-course dinner given by Pavlov's son that finished at 4 A.M.[59] After a morning of talks on 15 August, the Congress was transported about 30 miles to Detskoye (Tsarskoye) Selo, where Nicholas II and his family had lived in the Alexander Palace. The members were allowed to visit the apartments, untouched since the royals had been taken away in 1917. Many of the Russian physiologists appeared overcome at what they saw. From there, they walked across a great courtyard to the Catherine Palace, where a banquet for about 2,000 guests was held in the baroque Great Hall, with its intricate gilt mirrors and magnificent chandeliers. Polly noted, "It was the most astonishing scene of splendour, even though most people had got their ordinary, everyday clothes on." At the end of long tables were solid-ice animal models (mostly dogs in Pavlov's honor) "with caviare eyes and ribbons around their necks." There was also an ice model of Pavlov, carried in by procession, and he made a rousing speech in the middle of the dinner. Margaret noticed that during the three-hour dinner, the waiters were storing all the leftover food and wished that she could give them all her courses. After dinner, there was dancing and fireworks over the lake, so that the Hills arrived back in the city about 3 A.M. Polly was amused that this lavish entertainment merited just one line in the next day's newspaper: "A banquet was given in the Grand Throne Room of the Catherine Palace, arranged by the Leningrad Soviet and the Organizational Committee of the Congress."

Four special trains conveyed the Congress members to Moscow, arriving on the Saturday morning after overnight journey. On the train, AV was interviewed about Russian physiology for *Izvestia* by Marietta Shahinian, a well-known novelist and journalist. On arrival in Moscow they were met "by the inevitable fleet of Lincolns and taken along to the Hotel Metropole." The city immediately looked more prosperous than Leningrad. Hill and the International Committee visited the Kremlin to give their opinions to Molotov on the state of Soviet physiology. Hill recommended that they should take more foreign journals and publish some of their own work in languages other than Russian, as well as allowing young physiologists to gain experience in other countries. On the Saturday night, Molotov hosted a dinner for the Congress at the Kremlin and gave a long, inaudible speech surrounded by armed guards, to which Hill replied.

When they returned to the Hotel Metropole, a man tapped AV on the back. He turned to face Peter Kapitza, the physicist, who was living in the hotel after being refused permission to return to Cambridge in the summer of 1934. Over the previous decade, Kapitza established himself one of the leading experimental physicists at the Cavendish Laboratory, where he became almost a surrogate son to Rutherford. He told AV that his own movements were watched continuously, and he was frustrated by the authorities. Soon after Kapitza's detention, Hill wrote a letter to Pavlov about him that was hand-delivered by J.G. Crowther, the former Brigand and now science correspondent of the *Manchester Guardian*. Kapitza visited Pavlov in November 1934 and he offered the physicist a place in his lab, where Kapitza intended to "start experiments on the mechanism of muscles along lines similar to those of A.V. Hill."[60] Hill and Kapitza agreed to meet for dinner the next evening and were joined by Paul Dirac, the Cambridge theoretical physicist, who had been campaigning for his friend's release while spending the summer in Russia. After dinner, they all went for a walk so that they could talk confidentially; Hill noticed that they were being followed by two men in a car.

Two days later, the Hills arrived back in Leningrad where "without its Congress dress on [the city] looked more desolate than ever." Polly reflected: "It is potentially one of the most splendid cities in the world, and actually one of the most dismal." She concluded that "the only reliable sense-organ is the eye."

Notes

1. O.F. Meyerhof, "Über den Zusammenhang der Spaltungsvorgange mit der Atmung in der Zelle," *Berichte der deutschen chemischen Gesellschaft* 58 (1925): 991.
2. Needham (1971), 85–97.
3. Teich (1992), 229–231. It is now known that iodoacetate inhibits the glycolytic enzyme, *glyceraldehyde-phosphate dehydrogenase*.
4. Quoted in Peters (1954).
5. F. Lipmann, "Einar Lundsgaard," *Science* 164 (1969): 246–247.
6. Teich (1992), 229–231.
7. Teich (1992), 231.
8. E. Lundsgaard to A.V. Hill (11/1/31) AVHL II 4/56.
9. A.V. Hill, "The Revolution in Muscle Physiology," *Physiological Reviews* 12 (1931): 56–67.
10. Needham (1971), 98–105.
11. K. Lohmann, "On the Enzymatic Splitting of Creatinephosphoric Acid; at the Same Time a Contribution to the Chemistry of Muscle Contraction" (1934), translated by Teich (1992), 239–241.
12. While muscle activity certainly increases the demand for ATP, the synthesis of proteins from amino acids provides the largest steady consumption. It has been calculated that a typical cell in the human body uses 100 million molecules of ATP per second at rest. A. Flamholz, R. Phillips, and R. Milo, "The Quantified Cell," *Molecular Biology of the Cell* 25, no. 22 (2014): 3497–3500
13. In 1929, Carl and Gerti Cori, a husband and wife team working in St. Louis, showed that much of the lactate produced is exported from muscle cells and carried in the bloodstream to the liver, where it is reconverted to glucose. The glucose can then be carried back to muscle where it is stored as glycogen.
14. "Otto Meyerhof and the Physiology Institute: The Birth of Modern Biochemistry." NobelPrize.org. Nobel Media AB 2020, nobelprize.org/prizes/themes/otto-meyerhof-and-the-physiology-institute-the-birth-of-modern-biochemistry (accessed 23/3/2020).

Though their names are forever linked, Meyerhof, for some reason, disliked Embden, and avoided direct collaboration with him.

15. Needham (1971), 518–523.
16. C.L. Gemmill, "Recollections of Professor Otto Meyerhof," *Medical College of Virginia Quarterly* 2, no. 2 (1966): 141–142.
17. Hofmann et al. (2012).
18. T.P. Feng, "Looking Back, Looking Forward," *Annual Review of Neuroscience* 11 (1988): 1–12. Feng (1907–1995) was one of the founders of modern biological science in China.
19. L.E. Bayliss and P. Eggleton, "Fourteenth International Physiology Congress," *Nature* 130 (1932): 705–707. Leonard Bayliss (1900–64) was the son of one UCL physiologist, William Bayliss, and the nephew of another, Ernest Starling. After Philip and Grace Eggleton divorced, she married Leonard in 1939!
20. Angier II (1978):36.
21. Gemmill (1966).
22. P. Ball, *Serving the Reich* (London: Vintage, 2014), 43.
23. W. Lanouette, *Genius in the Shadows* (Chicago: University of Chicago Press, 1992), 116–120.
24. Lanouette (1992), 127. Within weeks of that meeting Szilard had the idea of a nuclear chain reaction, which caused him to abandon his interest in biology until after the war.
25. Correspondence on Society for Protection of Science and Learning AVHL II 4/78.
26. W. Beveridge, *A Defence of Free Learning* (Oxford: Oxford University Press, 1959), 1–6.
27. P. Weindling, "From Refugee Assistance to Freedom of Learning: The Strategic Vision of A.V. Hill 1933–1964," in S. Marks, P. Weindling, and L. Wintour (eds.), *In Defense of Learning: The Plight, Persecution and Placement of Academic Refugees, 1933–1980s* (Oxford: British Academy, 2011), 59–76.
28. D. Zimmerman, "Protests Butter No Parsnips: Lord Beveridge and the Rescue of Refugee Academics from Europe, 1933–1938," in Marks et al. (2011), 29–44.
29. Sir Walter Adams (1906–1975) became Secretary of LSE in 1938 when Hill wrote in a reference: "His sympathy for the exiles, his willingness to shoulder the burdens of others, and his wisdom in dealing with difficult personal and administrative problems, are combined with a critical and unsentimental judgment, and have made his services of quite exceptional value." He returned as Director of LSE during the student unrest of the late 1960s.
30. Lanouette (1992), 121.
31. A. Robinson, *Einstein on the Run* (New Haven, CT: Yale University Press, 2020).
32. Robinson (2020), 256.
33. Robinson (2020), 261–272. There is a video excerpt of Einstein's address available: royalalberthall. com/about-the-hall/news/2013/october/3-october-1933-albert-einstein-speaks-at-the-hall.
34. Commander Locker-Lampson's previous engagement at the Royal Albert Hall was with the "Blue Shirts," a short-lived anti-communist group. Robinson (2020), 201.
35. A.V. Hill, "The International Status and Obligations of Science," in Hill (1960), 205–221.
36. Ball (2014), 72.
37. Ball (2014), 88–97.
38. J. Stark, "International Status and Obligations of Science," *Nature* 133 (24/2/34): 290.
39. D. Wilson, *Rutherford: Simple Genius* (London: Hodder and Stoughton, 1983), 488–489.
40. R. Gregory to A.V. Hill (26&27/3//34) AVHL I 4/13.
41. J. Stark, "The Attitude of the German Government towards Science," *Nature* 133 (21/4/34): 614.

42. J.B.S. Haldane, *Nature* 133 (1934): 726. JBS was the son of Hill's old friend, J.S. Haldane, and became professor of genetics at UCL in 1933.

43. B. Sakmann, "Bernard Katz (1911–2003)," *Biographical Memoirs of Fellows of the Royal Society* 53 (2007): 185–202.

44. Katz (1978).

45. A.V. Hill, "Retrospective Sympathetic Affection" M&R 598–603.

46. G.V.R. Born and P. Banks, "Hugh Blaschko (1900–1993)," *Biographical Memoirs of Fellows of the Royal Society* 42 (1996): 41–60.

47. Weindling (2011).

48. Sir Hans Krebs (1900–1981) won a Nobel Prize in 1953 for his elucidation of the citric acid (Krebs) cycle that represents the ultimate biochemical pathway of oxidation. It is fed by the products of glycolysis (as well as by protein and fat breakdown), releasing carbon dioxide and energy that is used to generate the bulk of ATP.

49. J. Hill, "For Science, War or Humanity? A.V. Hill and the Academic Assistance Council" (unpublished dissertation). Sir Ernst Chain (1906–1979) shared the 1945 Nobel Prize for Medicine or Physiology for his work on penicillin. One obituarist noted that Chain showed "little patience with people he considered wrong or stupid ... and was not given to compromise." Hans Krebs in the mid-1930s regarded Chain as "a third-class chemist but a first-class violinist!" See Marks et al. (2011), 5.

50. J. Hill, "For Science, War or Humanity?"

51. A.V. Hill to W. Beveridge (1/1/34) quoted in D. Zimmerman, "The Society for the Protection of Science and Learning and the Politicization of British Science in the 1930s," *Minerva* 44 (2006): 25–45.

52. Weindling (2011).

53. www.cara.ngo/who-we-are/our-history (accessed 19/4/20).

54. Hill, Gowland Hopkins, and Kenyon quoted in N. Bentwich, *The Rescue and Achievement of Refugee Scholars* (The Hague: Martinus Nijhoff, 1953), 6.

55. P. Hill, "XVth International Congress of Physiology, August 1935" AVHL II 5/27.

56. W.O. Fenn (1968), 314–319.

57. J. Muckle, "Saving the Russian Children: Materials in the Archive of Save the Children Fund Relating to Eastern Europe in 1920–1923," *Slavonic and East European Review* 68, no. 3 (1990): 507–511.

58. P. Hill (1935). Meyerhof did not attend the Congress.

59. Fenn (1968): 317.

60. Crowther (1970), 139–140 and J.W. Boag, P.E. Rubinin, and D. Shoenberg (eds.) *Kapitza in Cambridge and Moscow* (Amsterdam: North-Holland, 1990), 211–212.

9

Preparing for War

During a speech winding up a House of Commons debate on disarmament in November 1932, Stanley Baldwin, the former prime minister and then a Cabinet member in the National Government, made the memorable prediction: "The bomber will always get through."[1] Often overlooked is his logic for saying so. Baldwin understood the vast volume of the skies, where a bomber might be concealed in cubic miles of cloud, with little chance of being spotted and downed by a fighter plane or anti-aircraft fire. The speed of bombers meant that British cities would be vulnerable within minutes of raiders crossing the coastline. As chairman of the Committee of Imperial Defence (CID), he admitted that he was depressed by "the perfectly futile attempts" made so far by the British government and internationally to deal with the prospect of aerial bombardment of civilian populations. Baldwin was worried that Germany would rearm and develop an air force, but he remained committed to the World Disarmament Conference as well as having faith in the League of Nations. Hitler withdrew Germany from both those institutions less than one year after Baldwin's speech. Winston Churchill sounded the alarm from the backbenches on 7 February 1934 that "this cursed, hellish invention and development of war from the air" nullified any defensive advantage from being an island nation, adding, "the crash of bombs exploding in London ... will apprise us of any inadequacy which has been permitted in our aerial defences."[2]

That summer the RAF carried out air exercises, mostly at night, over London; only about 40 percent of bombers were intercepted. Two of the targets were the Air Ministry and the Houses of Parliament: in the words of an observing journalist, "neither ... should bother any of us anymore."[3] Churchill kept pressing for expansion of the RAF and discussed the issue constantly with one of his closest friends, the Oxford physicist Frederick Lindemann. Lindemann, who was born and largely educated in Germany, was of the same generation as Hill and Meyerhof. He came to England in 1914 and, like Henry Tizard, taught himself to fly while working at the Royal Aircraft Factory at Farnborough. Lindemann worked out a theoretical solution to the fatal problem of uncontrolled aircraft spins and then courageously demonstrated his solution—saving many pilot lives. He came from a wealthy background and was an unabashed elitist and eugenicist, who despised "the fetish of equality." He was a brilliant expounder of science, although he produced little original research.[4] In the Easter vacation of 1933, his chauffeur drove him around Germany in his Rolls Royce in order to recruit scientists to come to the Clarendon Laboratory in Oxford, having first arranged some funding through Imperial Chemical Industries. He was not involved with the AAC. The handful of men he chose became the nucleus of a brilliant,

Bound by Muscle. Andrew Brown, Oxford University Press. © Oxford University Press 2023.
DOI: 10.1093/oso/9780197582633.003.0009

low-temperature physics group in Oxford. "The Prof" brought a new facet of analysis, by no means flawless, to the innumerate Churchill. In many ways, their friendship was the attraction of opposites: Lindemann was ascetic, non-smoking, sarcastic, tee-total, and a fastidious eater.

The Times published a letter from Professor Lindemann on 8 August 1934 under the heading "Science and Air Bombing."[5] Reflecting that all sides in a recent House of Commons debate had taken for granted that there was "no means of preventing hostile bombers from depositing their loads of explosives, incendiary materials, gases or bacteria" on their chosen targets, the Prof instead argued the defeatist assumption that no effective defensive method could be devised appeared to him "to be profoundly improbable." He urged the "whole weight and influence of the Government" should be engaged in the endeavor until "all the resources of science and invention have been exhausted." A few weeks later, Churchill and Lindemann were touring the South of France and made a detour to Aix-les-Bains, where Baldwin was on vacation, in order "to urge him to set up an air research committee to prevent German bombers from getting through to London."[6]

There was a two-man department of scientific research within the Air Ministry: A.P. "Jimmy" Rowe was the first assistant. He decided it was time to review all the material available pertaining to air defense and found just fifty-three files to study. Having read them Rowe wrote a memo to his director, Harry Wimperis, warning that "unless science evolved some new method of air defence," Great Britain was "likely to lose the next war if it started within ten years."[7] Wimperis, an engineer, had been in post for about a decade; during that period there had been several suggestions (mostly from cranks and charlatans) about employing "death-rays" against planes and their pilots. Sensing the new atmosphere of increased concern, Wimperis decided he needed to review the idea of death-rays and other innovations systematically. He decided to start by talking to the biologist who had unmatched experience in anti-aircraft defense: A.V. Hill. They lunched at the Athenaeum club on 15 October; no doubt Hill was skeptical about the prospect of using radio-frequency waves to disable or kill a pilot from a distance. He pointed out that for the purposes of radio-frequency heating, the human brain approximates one liter of water. Wimperis recorded in his diary: "I wanted a good talk with him on radiant energy as a means of A.A. Defence. I think I must put up a proposition to the Air Council."[8]

Wimperis made his proposition to Lord Londonderry,[9] the Secretary of State for Air, on 12 November, in which he referred to his discussion "on the physiological aspect" of death-rays with Hill. The most important recommendation that Wimperis made, and which Londonderry immediately accepted, was to establish an air defense committee under the chairmanship of Henry Tizard (Figure 9.1). Tizard, whom Hill first met at Upavon airfield in 1915, did not return to a traditional academic career after the 1914–1918 war, instead working as senior science adviser to government departments for a decade, before becoming Rector of Imperial College, London. Wimperis and Rowe would represent the Air Ministry, with the latter acting as the committee's secretary. Wimperis also suggested two independent scientists be appointed to the

Figure 9.1 Sir Henry Tizard (1885–1959)
Photograph by Howard Coster (1942) courtesy of the
National Portrait Gallery, London.

Tizard Committee (as it quickly became known): Hill, and Patrick Blackett. Blackett
was a brilliant physicist, who trained under Rutherford at the Cavendish Laboratory;
he had recently become an FRS and left Cambridge to be professor at Birkbeck College
London. Unlike Hill, he was a man of fixed political opinions—a man of the left—but,
having become a naval cadet at thirteen years of age and serving as a midshipman on
a battleship at Jutland when he was just eighteen, he understood war and the workings
of the military mind as well as Hill did. Tizard approached Hill after a council meeting
of the Royal Society later in November to ask him if he would serve. Hill replied he
would, providing they would be given access to all the information they might need,
and that secrecy would not be "used as a cover for incompetence."[10] The two men knew
each other well, and neither liked dissembling.

After his lunch with Hill, Wimperis approached Robert Watson-Watt, who ran the
government's Radio Research Station at Slough, where the main topics of interest were
the ionosphere and meteorology. Following up on Hill's idea to treat a pilot's brain as
a liter of water, Wimperis asked Watson-Watt about generating the necessary amount
of energy to cause radiofrequency heating of that much fluid from large distances. The

task fell to a young physicist, A.F. "Skip" Wilkins, who quickly showed it would be an impossibly enormous quantity. During his calculations, though, Wilkins realized that an aircraft would reflect enough electromagnetic energy to be detectable as a signal on an oscilloscope. Watson-Watt wrote to Wimperis, concluding, "attention is being turned to the still difficult but less unpromising problem of radio-detection as opposed to radio-destruction, and numerical considerations on the methods of detection by reflected radio waves will be submitted if required."[11] Wimperis had this reply in hand by the time the first meeting of the Tizard Committee was held in the Air Ministry at the end of January 1935. Two weeks later, he received a seminal paper from Watson-Watt on "Detection and Location of Aircraft by Radio Methods," with the author commenting:

> It turns out so favourably that I am still nervous as to whether we have not got a power of ten wrong, but even that would not be fatal. I have therefore thought it desirable to send you the memorandum immediately rather than to wait for close rechecking.[12]

Wimperis presented the memo to the Tizard Committee at their next meeting on 21 February, and they decided there should be a test without delay. Five days later, Watson-Watt, Wilkins, and Rowe found themselves in a field near Daventry, sitting in a converted ambulance that housed an oscilloscope connected to a simple wire antenna. They were waiting for a Heyford bomber to fly by on a prearranged route that would take it close to a nearby BBC radio mast. As the biplane approached, it disturbed the short-wave BBC signal and caused a trace on the oscilloscope screen when it was still eight miles away.[13]

Watson-Watt was a shy Scotsman whose prolix manner of speaking could be frustrating. Yet his memo, "Detection and Location of Aircraft by Radio Methods," conveyed in concise detail to the Tizard Committee a clear plan for an air defense system based on radio waves. There were technical specifications about oscilloscopes, wavelengths, and the need to identify friendly aircraft, as well as organizational proposals for several observing stations that would send data to a central control room, where the tracks of hostile aircraft could be plotted. He also suggested that a long spit of land on the coast of Suffolk, Orfordness, would be a suitably isolated site for further secret research. Tizard knew the airstrip there from his flying days and approved. A small group at Slough scrounged equipment from the National Physical Laboratory and from navy junk stores before transferring to Orfordness in May. If asked, they said they were carrying out "ionospheric research" with RDF equipment (RDF standing for radio direction-finding, soon to be known as *radar*). In mid-June, the Tizard Committee made a site visit, meeting in the Jolly Sailor pub near the quay at Orford. In the afternoon, Watson-Watt demonstrated his team's ability to track an aircraft flying off the coast during a thunderstorm. Wimperis wrote in his diary, "Got some lovely records up to 27 km."[14]

The affable, if not quite jolly, scientists of the Tizard Committee found that their harmonious existence was suddenly under grave threat at their next meeting on 25

July 1935. Professor Lindemann had been attempting to insert himself into the heart of air defense for many months by engaging Baldwin (now prime minister again), Lord Londonderry, as well as his political intimate, Churchill.[15] In doing so, he sought to belittle the Tizard Committee and succeeded in launching a new air defense sub-committee of the CID. The complacent Lord Londonderry was forced to resign as Secretary for Air and replaced by Lord Swinton, who chaired the new CID committee (which soon took his name). Appropriately he invited Churchill, the leading voice warning of the danger posed by the burgeoning Luftwaffe, to join. Churchill agreed on condition that Prof. Lindemann should be added to the Tizard Committee, which would continue to provide technical advice.

Tizard and Lindemann first met in Berlin in 1908, when they were both working in the physical chemistry laboratory of Walther Nernst. Tizard had been instrumental in Lindemann's securing his chair at Oxford in 1919, and they had remained on friendly terms. Yet, as Tizard's biographer noted: "Animosity between Lindemann and the other members . . . developed from his first attendance." It was not just a consequence of the Prof's caustic manner, although his belligerence certainly took a toll. There was a major difference of opinion on the potential of radar research, which Lindemann regarded as just one avenue that could be explored, whereas the rest of the Tizard Committee were convinced that it was in a class of its own. Lindemann gained no support for his unrealistic alternative schemes, notably suspending small mines of high explosive from parachutes in the flight path of hostile bombers. Lindemann was also used to operating with no delineation between his scientific and his political convictions, a tendency that the rest of the committee regarded as improper if not dangerous. Fortunately for the country, the Swinton Committee was sufficiently impressed by the progress being made by Watson-Watt and his team that it agreed to an expanded research effort and requested funding from the Treasury for a chain of radar stations to be built along the British coastline. Within three months, approval was granted to build the first stations between Kent and Suffolk—Chain Home—to protect the approaches to London.[16]

Hill and Blackett were not directly involved in the radar research but were valued for their experience as seasoned experimenters of proven judgment—scientists who would offer valuable advice and, when necessary, constructive criticism. Hill was also involved with physiological research for the Air Ministry, considering factors such as fog, moonlight, polaroid searchlights, dazzle, spectacles, and camouflage that might affect pilots' and A.A. gunners' vision.[17] In June 1936, Tizard wrote to Hill thanking him for his moral support and explaining:

> At any personal inconvenience, and in spite of any personal annoyance, I should go on trying to work with Lindemann if by doing so I thought the job would get on better. But I don't think it would: there would be continual friction. I did not want him to come on the committee as I foresaw this kind of trouble—but Swinton pressed it, so naturally we had to give it a fair trial; and I think we have given it a very fair trial.[18]

Tizard mentioned that he had written to Lindemann and was awaiting his response. The latest contentious issue between the pair concerned "wild criticisms" from Churchill, "presumably based on information from you."[19] Lindemann was not going to back down, accusing Tizard of being dilatory and complacent. The culmination came at the next committee meeting on 15 July with a shouting match so heated that the secretaries were asked to leave.[20] In the immediate aftermath of that row, both Blackett and Hill decided to resign. Hill wrote to Lord Swinton giving his reasons:

> Instead of being frank and open with his colleagues & the Chairman he went behind their backs & adopted methods of pushing his own opinions which—apart from anything else—would make further cooperation with him very difficult.[21]

Swinton replied to Hill the next day, very much hoping that he would continue in air defense work. He summoned Hill and Blackett for discussions in July, but they would not be tempted back. Over the summer Swinton dissolved the Tizard Committee, before reconstituting it in the autumn, with Edward Appleton, a physicist who had studied the ionosphere and radio waves extensively, replacing Lindemann.

Years later, Hill wrote to Lindemann's biographer, attributing the crisis in the Tizard Committee to:

> Lindemann's continual and violent advocacy of a fantastic scheme for dropping bombs, hanging by wires on parachutes in the path of attacking aircraft. When he started this wildcat scheme, he didn't believe in RDF and did his best to depreciate it. Then when he realized it might be a war-winner he tried to put Watson-Watt against Tizard; which was easy because Watson-Watt was readily led to believe that Tizard didn't appreciate him enough (indeed nobody could, except Watson-Watt himself). But Lindemann continued with his wild scheme and became ruder, more objectionable and less cooperative on the Committee.[22]

In Hill's considered opinion, the rest of the Tizard Committee was just as concerned about air defense as Lindemann was, just "not so passionate and unreasonable."

There was, in fact, an unrecognized but major difference of substance between Lindemann and Tizard. Lindemann believed the RAF could deal with bombers arriving in daylight, but he was far more worried about the deficiency of night defense. He noted that night defense was "more important than a cure for cancer."[23] Tizard, on the other hand, was trying to build an operational system starting from the most favorable conditions. Lindemann wrote Tizard an eight-page letter on the issue of night bombers in June 1935.[24] While radar beams obviously work in the dark, the Chain Home radar stations could only guide defending fighter pilots to within about five miles of an attacker, after which they would have to rely on visual spotting. At night, depending on the conditions, visibility would be restricted to 2,000 feet at best. The Tizard Committee began to address the issue in the summer of 1935; a conference mostly devoted to visual detection of aircraft from aircraft was held in

October with Hill in the chair. A Visual Detection subcommittee, also chaired by Hill, began regular meetings in February 1936 and developed a major project named "*Silhouette.*"[25] It was conceived as an array of upward-facing, wide-angle floodlights that would illuminate the cloud base. Fighters flying above the cloud would then be able to identify the silhouette of a bomber flying at a lower altitude as it moved across the ground-glass of the cloud, and they could dive to attack. Field tests were carried out at RAF Farnborough, and by the end of the year, Hill became the most enthusiastic proponent of *Silhouette*. Indeed, he suggested a larger examination, with an array of fifty floodlight stations spread across 200 square miles in Essex. The RAF agreed to support the expanded scheme, and the Tizard Committee was quickly able to secure 400,000 pounds (over £27 million in 2020 money) from the Treasury. The cost of installing such a system to cover major cities was just one reason for its demise by 1938. The installation of air interception radar sets in fighters, once feasible, was more economic and efficient, and two of the most senior RAF commanders Hugh Dowding and William Sholto Douglas were unimpressed by *Silhouette*.[26] The practical difficulty of looking down from a monoplane, with the wings in the way, was not lost on them.

As we have seen, in the mid-1930s Otto Meyerhof was consumed by his research program and making fundamental biochemical discoveries in the biochemistry, especially in the energy transactions mediated by glycolysis and ATP. But even the most blinkered and preoccupied scientist in Germany could not remain unaware of the inimical effect of the Nazis on his work. For a Jew, the brown-shirted savagery in the streets was an even more obvious threat. Walter Meyerhof was hit in the face by his music teacher at school. When his parents complained to the principal, he told them there was nothing he could do because the teacher was a Nazi Party member. Otto Meyerhof, like many decent German academics, at first believed that Hitler and his malign regime would be just an appalling but temporary aberration.[27] He seemed unperturbed enough to keep making meticulous observations of transient biochemical reactions, identifying the esterification of glucose that results in the formation of various phosphates, such as glucose-6-phosphate, that are the intermediary substances in glycolysis. Some of his younger colleagues were more perspicacious. Blaschko was one of the first to arrive in England to be supported by the AAC, and David Nachmansohn, after forging a successful career in Berlin-Dahlem, left for Paris in 1933. Meyerhof's position as head of a superb laboratory, built to his own specifications, added weight to the forces that kept him in his home country.

Bertolt Brecht, the playwright who fled Germany the day after the Reichstag fire in February 1933, later wrote a speech for a scientist character, Ziffel, who says:

> Working out whether you should flee today or whether you can leave it till tomorrow calls for the kind of intelligence which, just a few decades ago, might have produced an immortal work of art. Homeric courage is required to walk down the street; the self-denial of a Buddha is needed if you even want to be tolerated.[28]

Brecht gives voice to the existential torment of the persecuted. Although Meyerhof did not produce a work of art, he mustered enough spare intellectual capacity to reveal immortal truths about the chemistry of life.

The impact on German universities of the National Socialist German Workers' Party (NSDAP) party coming to power early in 1933 was direct and profound. On 5 April, the traditionally liberal state government of Baden anticipated the national mood by stating that all civil servants "of the Jewish race" would immediately be dismissed.[29] Two days later, the "Law for the Restoration of the Civil Service" was promulgated, at a stroke removing all persons of "non-Aryan descent" from their official positions. Since German universities were (and are) state-funded, the dismissals included academic staff; there was an exemption for Jews who had fought in the 1914–1918 war. The strategy of the NSDAP government was not to destroy the universities, but to suborn them into supporting the Nazi revolution in part by controlling the composition of the professorate and installing leaders, "Führers," who would ensure "coordination" with regime policies. Many of the professors at Heidelberg in 1933 were eager to fall in line, celebrating the demise of the Weimar Republic as un-German and weak, while praising Nazi policies despite their blatant discrimination against a significant proportion of their colleagues and their overriding, anti-intellectual bias.[30] There were also opportunistic careerists among the junior ranks, who could see that they might benefit from a shortcut through the often decades-long path to comfortable tenure. Even when there was little risk involved, there were few examples of resistance to the new regime from within academia. The most infamous Nazi professor at Heidelberg was Philipp Lenard, cofounder with Johannes Stark of the counterfeit "Aryan physics" movement. The university Rector, Willy Andreas, in his May 1933 commencement address, thundered:

> National Socialism has become Germany's destiny—it must fulfill its destiny.... More courageously and decisively than before, we shall place ourselves in the unleashed stream of the nation.[31]

Yet his praise for Hitler did not go far enough: some radical students mooted putting a bullet in his head. In July he was replaced as Rector by Wilhelm Groh, an undistinguished professor of labor law and member of the paramilitary Brownshirts.

Meyerhof occupied a hybrid position in Heidelberg: he was both an associate professor in the medical faculty—a civil service appointment—and the Director of Physiology in the KWG institute. Appointments at KWG institutes were made by the scientists themselves, not by the state, and they were partly funded by industry.[32] So Meyerhof's position was apparently protected by a number of factors—his eminence as a Nobel Laureate, his research position within the KWG (which also shielded him from abusive students), and perhaps his short military service—but in a university famous for dueling, these did not constitute a suit of armor. The directorship of the Institute for Medical Research was meant to rotate, and Meyerhof's turn should have come by April 1935. The Heidelberg NSDAP complained about this to the Baden leadership, saying it must be prevented. They were critical of von Krehl's failure to dismiss unsuitable employees (communists, social democrats, etc.), and his lack of control over Meyerhof. It was regrettable, in their view, that the documentary evidence against Meyerhof "unfortunately cannot be substantiated, because Meyerhof's staff refuses to bear witness against him under oath."[33] Richard Kuhn, a physical chemist and Nazi sympathizer who had fired all the Jews in his laboratory with alacrity, became the director instead of Meyerhof.

When he transferred from Berlin to Heidelberg, Meyerhof brought two laboratory technicians with him—Karl Schröder and Walter Möhle. Both were communists. Schröder was imprisoned in 1933 until Meyerhof managed to arrange his release. The KWG administration under political pressure decided that the pair should be fired in March 1934, but Meyerhof insisted that they were essential workers. Their dismissals were reversed on condition that he took full responsibility. Meyerhof also bent the rules of the KWG when it came to research staff. He offered an unpaid position to Hermann Lehmann, a young Jewish physician, who studied in Heidelberg but had to graduate in Basel.[34] Lehmann soon rewarded the laboratory with a series of excellent papers, including one that illuminated the reversible reaction between ATP and phosphocreatine and its dependence on pH. Lehmann was one of three non-Aryans given space by Meyerhof, a fact that infuriated the new director, Kuhn. He complained to Friedrich Glum, the director general of the KWG in Berlin, in April 1936:

> Allegedly there are again three persons of non-Aryan extraction working with Prof. Meyerhof in the institute (Mr. Lehmann, Miss Hirsch and another lady whom I don't know yet) a fact which will lead to discussions about the KW Society in general and the Heidelberg institute in particular. I suggest after investigating ... you give Prof. Meyerhof exact guidelines.[35]

Glum agreed that "Professor Meyerhof would be well advised to keep such cases to a minimum" but, because there were no actual employment contracts involved, the KWG administration decided not to concern itself in the matter.[36]

On the night in May 1935 when thousands of books were burned in Heidelberg, the student dueling corps "in full regalia, booted and sword-belted" marched in a torchlight procession alongside Nazi storm troopers.[37] Rector Groh, who had taken to wearing military uniform, warned that only professors committed to advancing the Nazi revolution belonged in Heidelberg. In 1935, the university decided to rename its physics department as the Philipp Lenard Institute. Hill came into possession of the Heidelberg students' magazine, surely via Meyerhof, in which Lenard responded to the warm congratulations of the student body. With reference to the continued preeminence of Einstein, Lenard warned:

> We must recognize that it is unworthy of a German—and indeed only harmful to him—to be the intellectual follower of a Jew. Natural science properly so called is of completely Aryan origin. ... Heil Hitler.

Hill forwarded the article to the editor of *Nature*, Sir Richard Gregory, who reproduced it in the "News and Views" section of the journal on 1 June. Gregory decided to publish it without comment in order, as he wrote to Hill, to "let it speak for itself in the scientific world."[38] Articles critical of the anti-Semitic and nationalist developments in universities under the Third Reich appeared regularly in *Nature*; by the end of 1937, subscriptions to the journal were banned in Germany.

This incident may have triggered Dr. Hermann Schlüter, the head of the NSDAP-controlled university teachers' organization, to write to Rector Groh, targeting Meyerhof:

> Professor Meyerhof is 100 percent Jewish and in the last few months has become increasingly active politically. There is evidence of not only scientific but also political contacts which he has with foreign countries. Therefore, Professor Meyerhof is highly dangerous, and an offer of an appointment from abroad offers an ideal opportunity to be rid of him. Despite his fame abroad, he is likely to vanish in the mass of other emigrants.[39]

The Nuremberg Laws promulgated in late 1935 meant that Otto and Hedwig were stripped of German citizenship and their right to vote. Their children, all raised as Lutherans to the extent that thirteen-year-old Walter had no idea he was Jewish, were all classified as Jews. The fig leaf of prior war service provided under the civil service law was now removed, and Meyerhof received a letter from Baden's minister for culture in mid-November informing him that "the question of the maintenance of your honorary professorship has now been decided in the light of yesterday's implementing regulation in the negative."[40] In January 1936, the *New York Times* carried an editorial pointing out that the fate of the two most eminent physiologists in the KWG, Warburg and Meyerhof, was "avowedly precarious."[41]

Heidelberg University used the occasion of its 550th anniversary in June 1936 to reaffirm its Nazi credentials. The Reich Minister of Propaganda, Josef Goebbels, who held a Doctorate in Philology from Heidelberg, gave the welcoming address.[42] Later that summer, Goebbels masterminded the propaganda coup of the Berlin Olympic Games and arranged for the temporary removal of anti-Semitic signs in the city.

A sports medicine conference was planned to coincide with the Berlin Olympics. Hill declined his invitation to attend, saying:

> I am sorry, for I have many German friends, but so long as the German government and people maintain their persecution of our Jewish and other colleagues, it will be altogether distasteful to me, as to most English scientists, to take part in any public scientific function in Germany.[43]

Other speakers at the Heidelberg celebrations emphasized the necessity of racial purity among scholars, the organic connection between academic work and the interests of the state and "Volk," whose racial hygiene must be protected by the medical profession, many of whom already recognized Hitler as the "teacher and adviser of every single German in all questions of the people's health."[44]

The increasingly hostile environment for Jews at least convinced Otto to make plans for his children to seek safety outside Germany. He arranged, through Hill, for Gottfried to complete schooling in England and then to study civil engineering at UCL—he graduated in 1938 and took a job with Ove Arup, the structural engineers. Hill was also instrumental in young Walter's traveling to England for a respite in the summer of 1936. Walter's sister, Bettina, returned to Three Corners for the month of August. Hill's colleague from UCL and SPSL, Charles Singer, welcomed a few Jewish children from Germany to spend their summers at his country house in Cornwall. As the end of the summer approached, Dorothea Singer asked Walter, "Do you really want to go back to Germany, where the Nazi boys at school will bully you probably even more than before? Why don't you go to school in England?"[45] Walter phoned his parents, who instantly agreed, given that his elder brother, Gottfried, and Otto's sister, Therese, already lived in London. He became a boarder at Dulwich College.

Ludolf von Krehl wrote to Max Planck, the KWG president, in October 1935 imploring him "most urgently to do everything in your power to ensure that Professor Meyerhof remains at our institute."[46] Planck did not address the issue directly, but Meyerhof wrote to Warburg in December 1935: "Herr Glum [the director general of the KWG] says my position here is untenable in the long run."[47] The KWG celebrated its twenty-fifth anniversary in January 1936 and was described in the Nazi party newspaper as "a playground for Catholics, Socialists and Jews."[48] In Lenard's opinion the KWS was "a Jewish monstrosity" conceived to enable Jews and their friends to secure "comfortable and influential positions as 'researchers.'"

It is remarkable given the purges and intrusion of the Nazi regime into university life that the years 1933–1935 marked the peak of Meyerhof's academic output, with over 100 papers published in conjunction with others in his department—nearly all written by Meyerhof himself. These were of foundational importance in biochemistry, especially his joint work with Kurt Lohmann on the breakdown and resynthesis of ATP in muscle. Meyerhof suggested there was an energetic coupling in cell metabolism between oxidation and phosphorylation. The fruitful collaboration with the apolitical Lohmann ended abruptly in 1937, when the authorities decided it was demeaning for him to remain in a department still headed by a Jew and created university chairs for him both in Heidelberg and Berlin. This promotion came despite the fact that Lohmann did not join the NSDAP, unlike most new biochemistry professors appointed after 1933.[49] Meyerhof was pleased for Lohmann's sake and wrote to Max Planck requesting that he be made a scientific fellow of the KWG.[50] In June 1937, Hill, as biological secretary of the Royal Society, seconded president W. H. Bragg's nomination of Otto Meyerhof himself to become a foreign member of the Royal Society.

In September 1937, Meyerhof finally decided to take steps to secure his own future. He traveled to the United States via London, with Hedwig and Bettina, hoping to arrange an academic position for himself and a medical school place for his daughter, who was a university student in Berlin. Although there had been some increase in research funding under the New Deal, the emphasis was on applied science such as agriculture that might lead to job creation; there were still many American scientists, who had lost academic positions in the early 1930s, looking for jobs. Incredibly, there was no university interest in one of the most accomplished and famous biochemists in the world. There was, however, a fateful meeting with one of his former researchers, David Nachmansohn, in New York. Nachmansohn, who moved to Paris with his family as soon as Hitler came to power, came to give a lecture at Yale, which ironically led to the offer of a job. He spent several days with the Meyerhofs and found Otto depressed by his own lack of success, having secured just one "offer of a minor and unsatisfactory position in an industrial concern."[51] Knowing of the admiration and respect for Meyerhof in French scientific circles, he offered to explore the possibility of a research professorship for him in Paris. Meyerhof was grateful and the two decided on a codeword that would indicate acceptance. On his return to Paris, Nachmansohn immediately contacted René Wurmser (biophysicist), Henri Laugier (Professor of Physiology at the Sorbonne), and Jean Perrin (Nobel Laureate in physics). They all lent enthusiastic support so that Nachmansohn was soon able to send Meyerhof a letter containing the prearranged codeword.

The 1938 International Congress in Physiology was due to be held in Zurich in August 1938, but there was considerable doubt about whether it would go ahead, especially after Hitler's annexation of Austria in March. Otto Meyerhof planned to attend the conference with Hedwig as a way of leaving Germany since his children were all out of the country. However, in March several German physiologists, including Meyerhof and Warburg, were refused permission to travel to Switzerland. Hill decided that he would support the conference and drove to Zurich with David, a physiology student, and Janet, a medical student. The German physiologists who did attend "had to conduct themselves at the meeting as a closed group with political guidance, subject to many internal regulations."[52] There were two further incidents that summer that made Meyerhof realize that he was in immediate danger. One was a message from the new president of KWG, Carl Bosch, that it was vital for him to leave Germany. And at the end of July, the Nazis forcibly liquidated the Meyerhof textile company in Berlin, imprisoning Otto's cousin, Justus, who ran the business, in the Sachsenhausen concentration camp.[53]

While the authorities would not permit Meyerhof to travel to Zurich, surprisingly they gave him permission to travel to Switzerland for a vacation. Otto, Hedwig, and Walter left Germany by train on 20 August, one day after the physiology congress finished. They left behind all their belongings. In mid-September, they took the train to Paris where his friends had arranged for Otto to become director of research at the Institut de Biologie Physico-chimique.[54] While his parents were settling in, Walter returned for one final term at Dulwich College. Meyerhof's departure from Heidelberg released laboratory space in the Institute of Physiology. This was eventually taken over by Richard Kuhn's chemistry group to research the modes of action of novel nerve gases such as Sarin for the German army.[55]

When Hill was trying to squeeze a pay raise for his instrument maker, A.C. Downing, out of the Foulerton Fund in 1930, he described him as "the best maker of galvanometers in the world."[56] By 1937, Downing had improved the thermopiles and galvanometers so that the system responded sixty times more rapidly than the setup that Hill and Hartree used in 1920. Using very thin specimens of frog sartorius muscle, Hill was now able to measure heat production at intervals of 0.2 seconds. In this way, he plotted the time course of heat production during isometric (constant length tightening) contraction. The muscle specimen was then allowed to shorten under different loads, in a maintained tetanic contraction, and Hill observed the release of extra energy in two forms: (a) "shortening heat" that is proportional to the shortening, plus (b) external work. The results represented an extension of the work Hill carried out after World War I in collaboration with Hartree, and more significantly a confirmation of the Fenn effect, discovered fifteen years before in the cold basement of Hill's Manchester house. For Hill, it at last removed any notion that muscle contraction could be explained by the rather passive visco-elastic model.[57]

In the fall of 1937, AV made an extensive tour of the United States, with Margaret and her companion, the recently widowed Norah Clegg, and this new work was the main subject of his lectures. Once back in London, AV and Margaret would resume their usual routine of meeting at the breakfast table, after he had taken his daily run and cold bath. Sometimes they were both home for dinner, although they were both taking on extra responsibilities that came to preoccupy them. AV would work late into the night, dictating papers, chapters, and letters. In addition to her role on Hornsey council, Margaret set up the Hornsey Housing Trust, with help from Norah Clegg and others, buying large houses to convert them into apartments for the elderly. She was appointed to a Royal Commission for the Location of Industry, which held its first meeting in October 1937, but she quickly lost patience with its turgid progress.

Although Hill had a burst of renewed enthusiasm for laboratory work in 1937–1938, his unremitting focus now was on the prospect of another war with Germany. He continued to attend meetings of the Tizard Committee. Tizard was overseeing close cooperation with the RAF to transform radar research into an operating defense system. As part of this, he was preparing universities to train physicists for radar work. Tizard was still tormented by the machinations of Churchill and Lindemann: he wrote to Hill in November 1938 that Churchill was "pressing Lindemann's claims to be the brains on defence."[58] By threatening to resign from the Swinton Committee on air defense, Churchill secured a place on it for Lindemann; Tizard managed to restore some balance by having Hill appointed too.

Orthodox military minds were turning toward the need for anti-aircraft batteries. There was a single A.A. division based near London under the charge of Major-General Frederick "Tim" Pile, who would become the head of a much larger A.A. Command once the war started. He wrote flatteringly to Hill in February 1938 that he had been told "you had forgotten more about A.A. than we had ever known."[59] Pile believed that "we lose the war unless we can hit enemy bombers from the first time they make their appearance. . . . In a nutshell, what I want is a method of practice without live ammunition applicable to units scattered all over London and if possible cheap." Pile added a postscript: "If you solve this, I have about a dozen other puzzles."

Hill regarded this as the moment the British Army first opened the door to operational research (OR). Pile invited Hill to the inevitable lunch at the Athenaeum, and this led to a close collaboration that both men valued immensely. In a handwritten dedication on an official report on the A.A. defense of the United Kingdom that he presented to Hill after the war, Pile wrote that Professor Hill "instigated the use of scientists in the field and whenever we were inclined to be puffed up told us a few scientific home truths."[60]

The workload of SPSL increased enormously after the annexation or Anschluss of Austria in March 1938 and with the deepening of fascism in Italy and Czechoslovakia. Hill had great faith in the abilities of Esther "Tess" Simpson, who ran organization with a remarkable combination of hardheaded efficiency and kindness.[61] For that reason, he was particularly responsive to her requests for assistance. At the end of 1938, Sir Richard Gregory asked Hill to write a report about the Society's work for *Nature*. Hill explained that 1,400 academics had been displaced in Germany alone, with more than 400 in Austria. The SPSL, through its disciplined efforts, had found permanent positions "for about 550 scholars in thirty-eight different countries, from Australia to Venezuela; for about 330 temporarily in twenty-five countries."[62] Hill went on to remind his readers: "Funds and interest are, however, an imperative need; first, for the work of the administration, information and advice; secondly, for direct help in human emergency." While Hill was the famous name leading the SPSL and emblematic of its fight against the fascist suppression of intellectual life, he also managed to take a personal interest in many of the scientists the Society supported. In February 1939, he hosted an evening conversazione at Burlington House, home of the Royal Society, where 300 leading British scientists mingled with 150 of their exiled European colleagues. He took as his guest a pretty seventeen-year-old girl from Vienna, Erica Guttmann, who had recently arrived in England by Kindertransport. Margaret had picked out her photograph and paid a fee for her to come to London. She was the Hill's house guest at Hurstbourne, and AV treated her with special kindness and interest.[63] He was a humanitarian who actually liked ordinary people.

The Royal Society increasingly became the center of his activities. In addition to routine committee duties, there were special moments to be enjoyed. One was when Sigmund Freud, who had been elected a foreign member in 1936, escaped to Hampstead from Vienna after the Anschluss. Freud, who had a painful jaw cancer, was not well enough to come to Burlington House to sign the charter book, so Hill, with two other Royal Society officers, took the priceless volume to Freud's house (Figure 9.2).[64]

Hill would also be involved with the honorary fellowships bestowed on King George VI and his prime ministers, Chamberlain, Churchill, and Attlee. Perhaps the diversion that brought him most pleasure was the research ship *Culver*.[65] The Royal Society Bermuda Oceanographical Committee agreed to provide a ship, in conjunction with the Woods Hole laboratory on Cape Cod, to be based in Bermuda to study

Figure 9.2 Sigmund Freud signs the Royal Society Charter Book, Hampstead 1938. (left to right) J.D. Griffith Davies (assistant secretary R.Soc.), Sir Albert Seward (foreign secretary R. Soc.), Sigmund Freud, A.V. Hill (biological secretary R. Soc.)
Photograph courtesy of the Wellcome Collection, London.

the gulf stream. An 80-ft. sailing ketch was purchased in 1937, with Hill signing the documents. She underwent an extensive refit for the role as a seagoing laboratory. As the nominal owner, AV decided that his youngest son, Maurice, who would be going to King's College Cambridge in October 1938 to read natural sciences, should join the crew. Maurice set sail on *Culver* from Plymouth in mid-June and reached Bermuda one month later.

By the summer of 1938, there was a feeling among senior university figures that there needed to be a systematic approach to ensure the greatest value from scientists and engineers in the setting of a new war. The Royal Society was seen as the agency best positioned to liaise with the government and to pull different science organizations together. The president, W.H. Bragg, discussed the issue with Hill and the new physics secretary, Alfred "Jack" Egerton. Like Lindemann and Tizard, Egerton had worked in Nernst's laboratory in Berlin before the Great War and remained friendly with both men. Egerton and Hill worked well in harness, both having equitable temperaments. They proposed to Bragg that a central bureau should be created in the Ministry of Labour to facilitate the optimal deployment of scientists and technical experts in the services or war industries.[66] Hill sent a memo to Bragg before a meeting at the ministry in December revealing a subtle approach:

> Much more than a mere card index of PhDs is needed. One thing is not to break up existing research institutions. Another is to make use of peculiar and special abilities; another is to have a proper set of categories; another to make tentative groupings of people likely to be able to help in particular problems.[67]

His reward was to be nominated as the official Royal Society representative on the Ministry's Advisory Committee.[68] Over the next several months, Hill and Egerton gathered details on physical scientists and less obvious candidates such as astronomers and naturalists. On one occasion, when Hill was chasing up data from a university department, he pointed out, "Hitler doesn't give one much time."[69] Two standardized cards, 8 in. × 5 in., were filed at the Central Registry for each scientist, containing information on age, address, qualifications, languages spoken, professional affiliations, career outline, nationality of parents, state of health, and military service.

By the outbreak of war in September, Egerton and Hill had collated details of about 7,000 scientists. This information was gathered on a voluntary basis, but in 1941 registration became compulsory for certain categories. One unhappy, retired, government scientist wrote a pompous letter of complaint under the pseudonym of "Ignotus" to *The Times*. Hill sent a blistering rebuttal the next day. After explaining that it was impossible, in an orderly system, to provide special forms for self-important individuals, he exulted:

> Sir William Bragg, O.M., and a former president, Sir Charles Sherrington, O.M., gladly filled up their two forms ... not because they can have supposed that they were unknown, but because their modesty demanded that they should be uniform with their colleagues. Sir William remarked to me that perhaps he might be able to teach physics to release a younger man.

Ignotus complained the cards were too small to accommodate his distinguished achievements, and Hill was pleased to inform him that "extensive experience with the Central Register has shown it is those of the least importance who write in the greatest detail of their qualifications."[70]

Hill was able to spend some of the last summer of peace at Three Corners and the MBS at Plymouth. In July, two young friends of David's from Cambridge, Alan Hodgkin and Andrew Huxley, were making their first electrical recordings from the giant nerve fibers of squid at the MBS. They were all staying at Three Corners, and Hodgkin described the scene in a letter to his mother:

> There was a rather holy atmosphere over the weekend as A.V. Hill came down bringing with him Sir William Bragg. A.V. is Secretary and Sir Wm. President of the Royal Society. They are both quite friendly and easy but all the same I felt that to have the President and Secretary of the Royal Society in one house was a bit too much of a good thing.[71]

In fact, AV had examined Hodgkin on his thesis, "The Electrical Basis of Nervous Conduction," at Cambridge in 1936, describing it as "a very remarkable piece of work for a candidate with—as yet—so short an experience in experimental research." Even before that he had encountered Hodgkin at the MBS in the summer of 1935, when he "realized then the high level of his intellectual qualities."[72] Hodgkin, for his part, was stimulated by a small monograph, *Chemical Wave Transmission in Nerve*, that Hill wrote in 1932 in which he argued that the membrane of a nerve cell is rendered permeable to potassium ions during activity.

In late August 1939, Hill attended a conference on adaptation to high altitude at the Scientific Station on Jungfraujoch in the Swiss alps. He was representing the Royal Society and took David and Maurice with him. Some German scientists had been permitted to attend; AV decided there was something seriously wrong "when all the Party members suddenly disappeared."[73] The Hills left immediately and, after phoning Margaret, drove for twenty-six hours to the French coast without stopping.

Maurice decided not to return to King's College and applied for a technical post in the Royal Navy. He was assigned to HMS *Osprey*, based on Portland Bill, where he was quickly immersed in research to improve ASDIC, the detection of submarines by reflected sound waves.[74] In November, Janet, by then a medical student, became the first of the Hill children to marry. Her husband, John Humphrey, was a friend of David's at Trinity College and a fellow medical student at University College Hospital.

Notes

1. *The Times* (11/11/32).
2. Roberts (2018), 376–377.
3. R.W. Clark, *Tizard* (Cambridge, MA: MIT Press,1965), 106.

4. R. Berman, "Lindemann in Physics," *Notes and Records of the Royal Society of London* 41 (1987): 181–189.
5. Clark (1965), 107–108.
6. Roberts (2018), 381.
7. Clark (1965), 107–109.
8. The Air Council was the governing body of the RAF.
9. Lord Londonderry (1878–1949) was a cousin of Churchill's, but nevertheless became a leading appeaser and admirer of Hitler's. See I. Kershaw, *Making Friends with Hitler* (London: Penguin Press, 2004).
10. Clark (1965), 112.
11. J.A. Ratcliffe, "Robert Alexander Watson-Watt (1892–1973)," *Biographical Memoirs of Fellows of the Royal Society* 21 (1975): 548–568.
12. Ratcliffe (1975).
13. blogs.royalsociety.org/history-of-science/2019/03/12/tracking-radar (accessed 12/5/20).
14. Clark (1965), 119.
15. Clark (1965), 120–127.
16. Clark (1965), 130–131.
17. Correspondence with the Air Ministry AVHL I 2/4.
18. H. Tizard to A.V. Hill (18/6/36) AVHL I 3/95.
19. Tizard-Lindemann correspondence reproduced in Clark (1965), 139–140.
20. P.M.S. Blackett, "Tizard and the Science of War," *Nature* 185 (1960): 647.
21. Clark (1965), 143–148.
22. A.V. Hill to Lord Birkenhead (19/9/60) AVHL II 4/10.
23. P.E. Judkins, "Making Vision into Power: Britain's Acquisition of the World's First Radar-Based Integrated Air Defence System 1935–1941" (unpublished PhD thesis, 2007), dspace.lib.cranfield.ac.uk/bitstream/1826/2577/1/Making%20Vision%20into%20Power%20-%20Phillip%20Edward%20Judkins.pdf (accessed 18/7/20).
24. Judkins (2007).
25. Judkins (2007).
26. Judkins (2007).
27. W. Meyerhof, www.youtube.com/watch?v=FZI_PeuY6LO.
28. B. Brecht, *Refugee Conversations* (London: Methuen 2019).
29. S.P. Remy, *The Heidelberg Myth: The Nazification and DeNazification of a German University* (Cambridge, MA: Harvard University Press, 2002), 15–16.
30. Remy (2002), 23.
31. Remy (2002), 23.
32. Ball (2013), 17.
33. Hoffmann et al. (2012).
34. J.V. Dacie, "Hermann Lehmann (1910–1985)," *Biographical Memoirs of Fellows of the Royal Society* 34 (1988): 406–449 (1988). Meyerhof was instrumental in arranging for Lehmann to transfer to Hopkins's biochemistry lab in Cambridge. In 1940, Hopkins and Hill secured Lehmann's release from internment as a "friendly alien." Lehmann became a leading expert on abnormal hemoglobin molecules.
35. R. Kuhn to F. Glum (27/4/36) quoted by Hoffmann et al. (2012). U. Deichmann, "The Expulsion of Jewish Chemists and Biochemists from Academia in Nazi Germany," *Perspectives on Science* 7 (1999): 1–86. Richard Kuhn (1900–1967) won the 1938 Nobel

Prize in Chemistry (awarded in 1939) for his work on "carotenoids and vitamins." He refused it because Hitler forbade Germans to accept Nobel prizes; in his letter to the Nobel Committee, Kuhn wrote that "The Fuhrer's will is our belief" (quoted in Deichmann 2002). He was allowed to accept the prize in 1949 despite his enthusiasm for Hitler.

36. F. Glum to R. Kuhn (29/4/36) quoted by Hoffmann et al. (2012).
37. S.H. Norwood, "Legitimating Nazism: Harvard and the Hitler Regime, 1933–37," *American Jewish History* 92, no. 2 (2004): 189–223.
38. A. Loewenstein, "Pragmatic and Dogmatic Physics: Antisemitism in Nature," in U. Charpa and U. Deichmann (eds.), *Jews and Sciences in German Contexts* (Tübingen: Mohr Siebeck, 2007), 23–44.
39. D. Meyerhof (2013), jewishcurrents.org/how-varian-fry-rescued-my-elders (accessed 7/6/20); Hoffman et al. (2012).
40. M. Emmerich, "Cells Flexing Their Muscles," www.mpg.de/7023399/S005_Flashback_086-087.pdf (accessed 8/3/20).
41. "The Last Stand," *New York Times* (13/1/36).
42. Norwood (2004). Charles Singer from UCL and SPSL wrote to the President of Harvard, James Conant, on 24 March 1936 in a vain attempt to persuade him to follow the example of British and European universities and boycott the event. Singer suggested that "the scandals at Heidelberg have been even above the normal level of German universities both in gravity and number."
43. J.A. Rall, "Nobel Laureate A.V. Hill and the Refugee Scholars, 1933–45," *Advances in Physiology Education* 41 (2017): 248–259.
44. Remy (2002), 58–63.
45. W. Meyerhof (2002), 31–32.
46. L. von Krehl to M. Planck (11/10/35) quoted in Hoffman et al. (2012).
47. O.F. Meyerhof to O. Warburg (17/12/35) in Werner (1996), 323.
48. Ball (2013), 96–97.
49. U. Deichmann, "Chemists and Biochemists in the National Socialist Era," *Angewandte Chemie International Edition* 41 (2002): 1310–1328
50. Hofmann et al. (2012).
51. Nachmansohn (1972).
52. W.O. Fenn (ed.) (1968).
53. www.discogs.com/artist/5539381-Ursula-van-Diemen (accessed 19/6/20). Justus was married to the opera singer Ursula van Diemen. He remained for months in Sachsenhausen until his family managed to arrange his release, after which was forced to leave the country. He came to London, where he committed suicide in 1944.
54. Hoffmann et al. (2012).
55. F. Schmaltz, "Chemical Weapons Research in National Socialism," in S. Heim, C. Sachse, M. Walker (eds.), *The Kaiser Wilhelm Society under National Socialism* (Cambridge: Cambridge University Press, 2009), 312–338.
56. Royal Society Collections CMB 64/39.
57. A.V. Hill, "The Heat of Shortening and the Dynamic Constants of Muscle," *Proceedings of the Royal Society of London (B)* 126 (1938): 136–195, and Rall (2014), 177–179.
58. Clark (1965), 178–179.
59. F.A. Pile to A.V. Hill (17/2/38) AVHL I 2/1.

60. F.A. Pile, "A.A. Defence of the UK, 1939–45." *Supplement, London Gazette* (18/12/47) AVHL I 3/77.

61. www.thejc.com/news/features/esther-simpson-the-unknown-heroine-1.438317 (accessed 6/4/20).

62. A.V. Hill, "Science and Learning in Distress," *Nature* 142 (1938): 1051–1052.

63. Weindling (2011). Erica stayed on in England after the war, and Professor Paul Weindling is her son.

64. There is a brief video recording of the occasion available at www.youtube.com/watch?v=pje-pzGILuc.

65. Royal Society research ship *Culver* AVHL II 4/18.

66. R.W. Clark, *The Rise of the Boffins* (London: Phoenix House, 1962), 55–60.

67. A.V. Hill to W.H. Bragg (18/12/38) AVHL I 2/1.

68. Minutes of Royal Society Council (12/1/39) AVHL I 2/1.

69. Clark (1962), 58.

70. Letters to *The Times* 22 and 23 May 1941 reproduced in A.V. Hill, "The Central Register of the Ministry of Labour and National Service," M&R: 332–336 CAC.

71. A.L. Hodgkin, "Beginning: Some Reminiscences of My Early Life," *Annual Review of Physiology* 45 (1983):1–16.

72. A.V. Hill, Notes on Hodgkin's thesis (3/9/36) AVHL II 4/36. Sir Alan Hodgkin OM (1914–1998) more than fulfilled AV's high expectations, sharing the 1963 Nobel Prize for Physiology or Medicine with fellow neurophysiologists Andrew Huxley and John Eccles, before becoming in turn president of the Royal Society and master of TCC.

73. Angier II (1978):107.

74. E.C. Bullard, "Maurice Neville Hill (1919–1966)." *Biographical Memoirs of Fellows of the Royal Society* 13 (1967): 192–203.

10
1940—Backs against the Wall

Without the support of the influential colleagues in Paris recruited by David Nachmansohn, the Meyerhof family would not have found it easy to flee to France in September 1938. Many countries, including France, were accepting very few alien Jews. The Meyerhofs were provided with a small apartment in the 5th Arrondissement—the epicenter of higher education in Paris. In early December, Otto took up his position as research director at the Institut de Biologie Physico-Chimique (IBPC). The previous month, a displaced young Polish Jew living rough in Paris had murdered a junior diplomat at the German embassy. This served as the pretext for the explosive pogrom across Germany: Kristallnacht. Days before Meyerhof took up his position at IBPC, he was officially dismissed as director of the KWI in Heidelberg, a consequence of the by-law of 12 November 1938, "on the exclusion of Jews from German business life."[1]

The IBPC is a multidisciplinary institute for the study of life sciences, established in 1931. Initially proposed by the physicist Jean Perrin and funded by the Jewish philanthropist, Baron Edmond de Rothschild, the IBPC is housed in a functional, brick building on the Rue Pierre et Marie Curie.[2] While Otto was able to re-engage in a busy routine, Hedwig could not disguise the anxiety and depression caused by the enforced changes in their circumstances. She would confide in sixteen-year-old Walter, now at school in the nearby Lycée Lavoisier, that she had been a bad wife and companion to Otto. Like most professional families, the Meyerhofs had always been used to having domestic staff and Hedwig, the artist and mathematician, never learned to cook. Walter was a sympathetic son and would put his arm around his mother to console her.[3] At weekends, the Meyerhofs and the Nachmansohns often visited cultural sites such as Chartres and the chateaux of the Loire, where Otto would reveal his deep love of architecture and history. These pleasant trips came to an end in the early summer of 1939, when the Nachmansohns left for Yale.

Meyerhof's laboratory facilities at IBPC were much more limited than in the department he helped to design in Heidelberg, yet the biggest losses were his faithful lab assistants and above all the presence of Karl Lohmann, now in Berlin. Nevertheless, he managed to publish a handful of papers, mostly on aspects of glycolysis as well as one about Goethe's scientific methods. He kept a keen eye on new publications in biochemistry. He wrote to Warburg in June to complain that his own contributions were

Bound by Muscle. Andrew Brown, Oxford University Press. © Oxford University Press 2023.
DOI: 10.1093/oso/9780197582633.003.0010

not being acknowledged. After referring to the constant worry about his relatives, he told Warburg:

> I am well taken care of here, can do what I want and my financial position is relatively good. Apart from lack of space, the chief problem is that the lack of trained and suitable personnel, together with a realisation of the necessary difficulties of equipment. Language and settling in, have forced me to start again *ab ovo*, like in Kiel in 1912. But I am not complaining, freedom is very precious.[4]

He did have one major complaint, however. Despite the exigencies of life in Heidelberg in 1937–1938, Meyerhof published a series of meticulous papers showing that during one stage of glycolysis, an oxidative reaction in muscle extract is linked to the phosphorylation of a molecule of ADP to ATP, in a strict one-to-one fashion that Meyerhof named "energetic coupling." This essential preliminary work carried by his group was being unjustly ignored in publications from the Warburg laboratory in Dahlem. As evidence that he was not just being paranoid, Meyerhof said that an unnamed Cambridge biochemist recently visited Paris and "brought up the subject himself in our discussion, that the lack of mention had been noticed in the laboratory there and where did I stand in regard to it." He pointed out that he had warned Warburg against such behavior before and now found himself being subjected to it. "I am writing to you with this frankness," he continued, "because I do not want our personal relationship to suffer now that we are, to use the historic phrase, 'standing on opposite sides of the barricade.' You will have heard that my Heidelberg Institute is now practically speaking dissolved and now partly occupied by Kuhn." Meyerhof could not know that Kuhn would soon be using his old physiology laboratory to research the mode of action of and possible antidotes to nerve gases of the Sarin type for the German army's ordnance department.[5]

On 2 September, the day before Great Britain and France declared war on Germany, Otto and Hedwig were exempted from the risk of internment as "persons of importance" thanks to interventions by Jean Perrin, Henri Laugier, and others with the Ministry of the Interior.[6] Otto wrote regular letters to Gottfried, his eldest son, who was working as a civil engineer in England. On 3 September, he wrote to Gottfried for the last time in "good German," preferring "bad English" to make life easier for the British censor.[7] The day after France declared war, notices went up around Paris instructing all German males between the ages of seventeen and fifty-five years to present themselves to a sports stadium for internment. Walter complied and was taken by bus to a camp at Moulins in central France.[8] His father contacted the local prefecture, who refused to release him without a direct order from the Ministry of the Interior. Meyerhof persuaded Perrin, rather against his will, to accompany him to the Ministry, where they were received by the chef du cabinet, "in the most elegant room."[9]

It soon transpired that Walter had been transferred to a different camp, further south, so that Otto returned to the Ministry on his own, a few days later. He made the case to an assistant that he was distracted from his scientific work for the national defense by his preoccupation with Walter. The Ministry gave the order for his release, and Walter arrived home after an absence of three weeks. In his letter to Gottfried, Otto suggested the British would be better at discriminating between genuine refugees and Nazi sympathizers. Otto thought Gottfried should be safe since he held all the correct documentation, but as an added insurance he had written separately to Hill. Otto closed by saying, "This war is a terrible thing and will last long. But I am confident of the victory of the democracies and the perishing of the Nazi hell."[10]

Walter had already passed the examinations to study physics at the École de Physique et de Chimie in Paris. Hedwig, especially, was relieved to have him home again in their tiny apartment, where Walter waited impatiently for classes to start in October. Gottfried was classified as a "friendly alien" in England with some restriction on his movements, but he was hoping that these would be lifted soon by a tribunal.[11]

Otto started to write letters to American academic institutions again, trying to secure a position. He discovered there was even less interest in his services than there had been in 1937. Bettina wrote regular letters from Johns Hopkins Medical School that arrived promptly by the clipper service. She had taken it upon herself to find a professorial position for her father in the United States, and recruited a powerful supporter in Abraham Flexner, who directed the new Institute for Advanced Study at Princeton where Einstein was in residence. Her father remained somewhat ambivalent. He wanted to keep his options open in case the situation in France "develops unfavorably," but with spring approaching he was inclined to say put, "sharing in a small way the common task and burden to knock out the German gangster leader."[12]

The spring of 1940 was an agreeable time to be in Paris, and the national mood was still confident. The Meyerhofs had identity cards that were valid for two years and had access to their French bank account. In April, they were invited to dinner with Baron Édouard Rothschild, the head of the family bank, at his grand house on the Place de la Concorde. Another guest was Marie Bonaparte, a Greek princess and psychoanalyst, who had extracted Freud from Vienna and installed him in her house in London (where Freud signed the Royal Society charter book). In early May, Meyerhof wrote to Hill mentioning that Hedwig was better, although still prone to situational depression.[13] She had her eyes set on the United States, where Bettina was very happy, but he regarded such a move as impossible. He was inclined to face the dangers in France and become fully assimilated there, although he did recognize that defeat in the war would mean they were lost. Days after writing this letter, Otto and Hedwig took a walking holiday in the Roche-Guyon area north of the city, but his confidence was shaken by news of new German military invasions. Otto wrote to Gottfried, "the not unexpected invasion of Hollande and Belgium aggravates our situation. The next days will show if the North of Hollande can resist and I hope to have news about the fate of our many relatives."[14] By the time he wrote this letter, the Germans had taken the French town of Sedan, south of the Belgian border, and were crossing the River Meuse into the Ardennes.

While the Meyerhofs were away, fresh notices were pasted on Parisian walls, instructing male and female Germans to report, bringing a blanket, clothing, and food for a few days, so Walter disappeared again.[15] On Otto's return to Paris, he told Gottfried he was hoping "after the fall of Bruxelles the front will stand firm and the valiant French army will withstand the avalanche of armoured cars and tanks."[16] He had been to see the chef du cabinet at the Ministry of Justice to request French naturalization to protect the family from internment. He warned Gottfried that they were under great tension and preparing to evacuate at short notice if necessary. The only comfort was that most of their financial assets were deposited in British and Canadian banks.

While the Ministry of Justice considered his request, Meyerhof was again able to call on his friend Perrin, who was the president of the High Committee for Scientific Research at the Ministry of Education. Perrin immediately issued him with a certificate stating that his petition of naturalization was being processed and confirming he was employed at IBPC on "work in the interest of the national defense."[17] Despite this, Hedwig was detained in a camp, but with the certificate and confirmation that they were being considered for naturalization, Otto obtained her release after twelve anxious hours. In a birthday letter to Gottfried, Otto wrote that he considered leaving for Bordeaux, but was informed that no scientific worker was allowed to leave Paris. His wishes were for his children's futures, trusting "you will be ready to fulfill all duties and efforts for the common good, to save Europe from Barbarism and decay; finally, we hope to unite once more after this terrible war is over."[18]

Gottfried spent an Easter vacation at Three Corners in Devon and was able to learn about some of AV's recent activities, although he was not present. During 1939, Hill was sounded out by his brother-in-law, Tom Hele, Master of Emmanuel College, to consider standing as parliamentary candidate for Cambridge University. The university had been returning MPs to the House of Commons since the early seventeenth century—previous representatives included such luminaries as Newton and Pitt the Younger.[19] In the modern era, Cambridge graduates were allowed to elect two MPs—nearly always Conservatives. When one of the sitting MPs died, expectedly, in December 1939, Hill was ready to throw his hat into the arena but was concerned about the party affiliation. Writing to a senior Conservative MP, Hill said he thought there would be a special advantage in selecting a scientist as the Cambridge University MP "particularly now" because "although science is called in for technical advice by almost every department of state, it has rather little influence in helping to guide policy. It is still the handmaiden and not a partner in government."[20] He explained that such concerns had been discussed extensively in the councils of the Royal Society over the past eighteen months, adding that the president, W.H. Bragg, would prefer him to stand as an independent. The selection committee of Cambridge graduates met at the House of Commons on 18 January and invited Hill to stand as an "Independent Conservative" candidate. They expected that "as an Independent supporter of the present government," Hill would undertake to vote with the government on major issues but otherwise he was free "to exercise your own unfettered judgments."[21] This gave Hill enough leeway to accept, and he drafted his first stump speech: "I have never been connected with a political party and have regretted, for example, in the past both the weakness, as it seemed to me, of Conservative foreign policy, and the short-sightedness of the Labour opposition to re-armament and civil defence."[22] The over-riding consideration now was the vigorous prosecution of the war, which Hill believed would be helped by bringing science and learning to national policy making. To introduce himself to the small electorate, which was widely disbursed, he listed his main current duties:

> 1   Secretary of the Royal Society
> 2   Chairman Scientific Research Committee, Central Register
> 3   Member of Committee for the Scientific Survey of Air Warfare [Tizard Committee]
> 4   Associate member of the Ordnance Board
> 5   Chairman of the Executive Committee of the National Physical Laboratory
> 6   Member University Grants Committee
> 7   Member Scientific Advisory Committee, the Ministry of Supply and Vice-Chairman Society for the Protection of Science and Learning

He did have an opponent at the by-election, John Ryle, a distinguished physician who was the Regius Professor of Physic at Cambridge, standing as an "Independent Progressive." The *Cambridge Review*[23] presented the candidates' answers to a short list of questions. The first was, simply, "Will you support the present government?" Hill answered "Yes," whereas Ryle, while recognizing no immediate change of government was possible, said he had "no confidence in the present government." He wanted "an announcement of peace-aims calculated to inform, inspire and unite a great democratic people." They gave similar views on the conditions necessary for an armistice (German withdrawal from Czechoslovakia and Poland, free elections in Austria, internationally guaranteed borders) and both supported the idea of a European federation post-war. While Ryle wished for immediate peace talks and refugee relief in central Europe, Hill discussed public health in the United Kingdom and warned, in an almost pre-Orwellian tone, of the need to "avoid excessive state control of independent persons and institutions." The by-election was held by postal vote in February, and Hill triumphed by an almost 2–1 margin.

Hill took the oath of allegiance to the Crown at the House of Commons on 27 February but would not take up his seat for several months (Figure 10.1). In November 1939, he had agreed to a request from Tizard to travel to Washington as a scientific adviser to the British air attaché. In the previous couple of years, there had been fruitless visits to the British Embassy in DC by Royal Navy and Air Force officers, who tended to be dismissive of American military technology.[24] The Navy believed any exchange of information would be to the advantage of the United States, while the Air Force was only really interested in the Norden gyroscopic bombsight that the Americans refused to transfer under any circumstances. Hill and Egerton, at the Royal Society, discussed "a dangerous lack of liaison"[25] with the scientific organizations in the Dominions, especially the National Research Council in Canada, which they regarded as a back door into the United States. They alerted government departments and the Canadian High Commissioner about their concern. On 20

Figure 10.1 Hill's first day as an MP for Cambridge University, Westminster February 1940

Professor Nick Humphrey/Churchill Archive Centre, Cambridge (Papers of A.V. Hill, AVHL).

November, Hill wrote optimistically to Tizard that: "It seems rather likely that scientific liaison will be established with Canada, and the American proposal should— apart from anything else—make that much more effective."[26] Tizard's ambition was to encourage American scientists into the war before their country. Hill had many personal contacts through his regular visits to the United States and spent the first weeks of 1940 asking colleagues for the names of other Americans who were known to be pro-British. They responded not only with the names of senior academics, but more importantly with the names of those critically placed in industry and government circles. His friend and former Brigand, Ralph Fowler, for example, urged Hill to meet Vannevar Bush in Washington, whom he described as "a man of wide interests."[27] Bush was already president of the Carnegie Institution and of the National Advisory Committee of Aeronautics (the precursor to NASA); in June 1940, he would persuade President Roosevelt to establish a National Defense Research Committee (NDRC) with himself as chairman. Hill collected names and thumbnail sketches on scraps of paper, the backs of Royal Society lunch menus and on notepaper from the *Lancastria*, an ocean liner converted to a troopship, in which he sailed from Liverpool on 9 March. The ship arrived in Halifax ten days later, after zigzagging across the North Atlantic; Hill immediately took a train to Washington.

As one might expect, the most influential connection he made was the British ambassador, Lord Lothian. A Liberal politician, Lothian had been appointed the previous summer because of his deep understanding of the American political system and his extensive network of influential friends. He first met President Roosevelt during the Paris peace conference in 1919 and his cousin was also an old friend of FDR's.[28] Hill wrote to his family in London a few days after arriving at the embassy. After saying he had not had time yet to form any political impressions, the one certainty in his opinion was "that America won't come into the war, or do anything very positive, until after the Presidential Election is over [November 1940]. . . . America is now in the funny indeterminate state of mind that we had been in for many years until the aftermath of Munich roused us. They are as ready as we were to let the Dictators gain one point after another by slow attrition, and never making up their minds."[29] He viewed journalists and politicians who opposed intervention, as magnifying British misdeeds over the last fifty years in order to have "salves for their consciences for not joining us in what they really all know is their cause." Democracy, he concluded, is rather slow in competition with dictators.

His first, informal progress report to Tizard, written the same day, was less plain-spoken: "I have met a lot of Generals and Admirals in the US Services and they seem peculiarly friendly. We chaff each other about secrets we should like to tell each and must not."[30] He said he was about to call on various universities in New York (to meet Leo Szilard and Enrico Fermi at Columbia), Boston (James Conant and Walter Cannon at Harvard), and Philadelphia, where he would attend the annual conference of the American Philosophical Society. He had also made lists of American Army and Navy officers, and organized names of engineers and physicists by the companies they worked for. He was already impressed with the availability of young engineers in the United States versus their scarcity in the United Kingdom. He was wondering about inveigling US academics to supplement British research programs with graduate students. The following day, Lothian arranged for Hill to have dinner with Felix Frankfurter, associate justice of the Supreme Court and a trusted adviser to the president.[31]

When he visited prominent electrical companies such as Sperry Corporation on Long Island and Bell Laboratories in New Jersey, he was accompanied by a US naval officer to stop him seeing things that he was not meant to; a request to visit the RCA research facility, also in New Jersey, was denied. The British officials in Washington did not have a clear idea about the state of radar research in the United States, and Hill's initial view was that American military officers did not seem to have any serious interest in the subject. Hill only had time to attend the last day of the APS conference on 20 April, when he gave an after-dinner speech on "Internationalism in Science," before returning to Washington to attend the annual meeting of the National Academy of Science that began on 22 April. Although RCA had not allowed Hill to see their microwave research facility, he was able to discuss the technology with Bush, Alfred Loomis, and Mel Tuve. He gained the impression that American radio tubes could generate far more power than the British counterpart.[32]

After his talks in Washington, he immediately dictated a letter of historic significance to Tizard. While the "Americans are on to a large number of applications of RDF [radar]," he said, "These are in the earlier development stage."[33] He still believed that the military did not seem to know much, but the research and development departments of the leading electrical engineering companies were working hard, especially with ultra-shortwave applications. His tone became emphatic: "We are clearly a long way ahead, particularly in experience under service conditions, but they will catch up fast and our best bargaining position probably exists NOW."

He suggested giving the Americans complete, detailed information so that with their big resources they could address specific remaining questions. He believed there was strong, if latent, goodwill toward the British war effort and warned "our bargaining power will steadily disappear" so that it would be "much better to be frank, generous and immediate."

Ambassador Lothian was convinced by Hill's case. They already knew from Justice Frankfurter that President Roosevelt was likely to support open technical exchanges. Lothian sent a cable to the Foreign Office, also on 23 April, that repeated much of Hill's argument to Tizard, but also dangled the idea of later trying to acquire the Norden bombsight, as well as suggesting the exchange of military and technical information would improve Anglo-American diplomatic relations generally. The next day Hill took the train to Schenectady in upstate New York, where the General Electric works was the largest electrical engineering plant in the world. GE had been investigating radar for years and were about to freeze all their commercial radio development to release engineers for defense-related radar research and to scale up production of equipment.[34] While he was on a train from Schenectady to Rochester, where he was going to visit Kodak, Hill wrote a few lines home that acutely summarized his achievement and concerns:

> I attended the National Academy of Sciences and did some rather useful work which resulted in two long cables to the Air Ministry and a cable by Lothian to the FO. I hope it will be useful but I fear our people will be too thick-headed.[35]

In fact, the communications from Washington were enthusiastically received in London, both at the Foreign Office and the Air Ministry. Lord Halifax, the Foreign Secretary, made a positive presentation of the recommendations to the War Cabinet on 26 April.[36] Tizard lobbied the Chief of the Air Staff, Sir Cyril Newall, and through him, the Secretary of State for Air. Tizard strongly recommended the frank interchange of all technical information with the US Army and Navy. He also suggested: "It must help materially to enlist the help and powerful resources of the American radio manufacturing industry," which would not be possible without disclosing technical secrets.[37] If the Air Ministry agreed, Tizard favored sending a special mission as soon as possible, but did not think "that Hill is the best man to convey the information and do the bargaining. He is not a radio expert, and Americans are such good salesmen that he may well run the risk of attaching too much importance to some of their claims." To Hill, Tizard merely wrote that he was "in favour of the interchange of information you suggest, but there is some difference of opinion about the policies now being discussed on a high plane. I will see that the people concerned get your most recent information."[38]

The tenth of May 1940 was a fateful day in World War II, on which Hitler oversaw the invasion of the Low Countries, and Churchill replaced Chamberlain as British prime minister. AV wrote home to "Marg & Co" that evening from his Washington hotel. After exclaiming "What a day!"[39] and sensing that the "phony war" was over, he welcomed the onset of large-scale fighting because he thought it would deplete Germany of oil and other resources. "The war can't be won without fighting"; he continued, "and, without winning, civilization in Europe is at an end." He sensed the events of the day were galvanizing his American colleagues. He had dinner that evening with the Blakeslee brothers (Albert, who was president of AAAS, and George, the head of the Brookings Institution), who both supported the early entry of the United States into the fray. His old friend, Detlev Bronk, phoned from Philadelphia and railed about "the damned shame" of America not joining in tomorrow. He asked AV to find him a war-related job. Hill stuck to his view that the United States would not enter the war prior to the November election, although "Churchill is popular here." If he could "form a decent Cabinet," Hill thought there would be a "better chance of cooperation and 'unneutrality.'" He supposed London would be bombed soon and, signing off as A.V.H., sent "Best love to you all and keep out of the way of the Blitzkrieg."

In March 1940, Tizard had received a typed memorandum from two physicists in Birmingham, Otto Frisch and Rudolf Peierls, who calculated the critical mass of pure uranium-235 that would be needed to support a chain reaction under bombardment by fast neutrons. Their findings suggested that an atom bomb could be constructed with pounds rather than tons of enriched uranium; other leading nuclear physicists in England could find no errors in their assumptions or calculations. As part of a British intelligence survey in April, Hill was asked to find out what the American attitude was toward possible uses of nuclear fission. After talking to Fermi, Bush, and others, Hill reported that they regarded any practical application of such scientific knowledge as "not inconceivable" but a long shot. He passed on their view that "it would be a sheer waste of time for people busy with urgent matters in England to turn to uranium as a war investigation." There were a large number of American physicists interested in the subject, who could be relied upon to keep the British informed if "anything likely to be of war value emerges"; in the meantime their research "for present practical needs [is] probably a wild goose chase."[40] This advice was given before either Hill or the American scientists knew about the Frisch-Peierls memorandum.

Hill made two visits to Canada, the first in April continuing his train trip from upstate New York, and the second in May. He wrote to Tizard saying he spent time with Banting in his laboratory in Toronto discussing aviation medicine. He confessed to being mystified as to why Don Solandt, a superb physiologist who had worked with him at UCL and was an experienced pilot, was not included in the research but felt he must not interfere. He told Tizard, "I have come to conclusion that if a successor to me were appointed when I return, he could better be posted to Ottawa."[41] He thought it would be just as easy to continue scientific contacts with Americans from there. His recommendation for the job was his old friend, Ralph Fowler, the first man to be appointed to a chair of theoretical physics at the Cavendish Laboratory. He would be especially qualified to liaise over the uranium-235 business. Fowler might have been considered a surprising nomination since he had experienced a severe stroke the previous year—writing to his protégé, Paul Dirac, Fowler felt as though he had "faded out."[42] He told Hill at the beginning of the war, "Others people are going to take risks now, so am I."[43] A letter of frustration to Tizard followed a few days later from Hill saying he had done all he could in Washington and wanted to come back to stir things up in London. He continued:

> My main thesis is that <u>we could get much more help in the US and Canada if we were not so damnably sticky and unimaginative</u>. There is an intense eagerness to help which we do not exploit: (a) because we are such bloody asses, or (b) because we are so sure we can win without anybody's help.[44]

Tizard replied that he had written to the Chief of the Air Staff recommending that Fowler be sent to Canada, but revealed that he had no idea what had happened about Hill's suggestion about giving secret information to the US government, even though he had supported it "very strongly" and urged that a special mission be sent.[45] The simple answer is that the calamitous setbacks in Europe, with the imminent collapse of France, changed the hand the British government was playing. Churchill had been lukewarm about the Hill-Lothian plan to begin with, skeptical of the idea that information about radar should be given up without gaining something in return. His reluctance was then increased by some military chiefs, who did not believe the Americans had anything to offer and mistrusted their security arrangements. An exception was Admiral of the Fleet, Sir Dudley Pound, who was supported by Churchill's successor as First Lord of the Admiralty, A.V. Alexander. They asked the new prime minister on 20 May to make a direct approach to Roosevelt regarding an "unrestricted offer to pool technical information" without appearing to "bargain."[46] Churchill responded that while he was intent on retaining the goodwill of the United States, he did not think "a wholesale offer of military secrets will count for much at the moment."[47] With the German army now rapidly approaching the French channel ports, Churchill's preoccupation was how to defend against an imminent invasion.

The immediate pressure was relieved by the improbable rescue of tens of thousands of men from Dunkirk, but Churchill was still desperate for military hardware from the United States. A few days after the evacuation of Dunkirk, Churchill revealed his displeasure with the lack of American support. Writing to the Canadian prime minister, William Mackenzie King, on 5 June, he complained:

> Although the President is our friend, no practical help has been forthcoming from the United States. We have not expected them to send military aid, but they have not even sent any worthy contribution in destroyers or planes, or by a visit of their fleet to Southern Irish ports.[48]

By this time Hill was steaming back to England on the *Scythia*, another Cunard liner converted into a troopship. The perilous state of Great Britain where his family were, and the fact his recommendations regarding collaboration with scientists in North America were being ignored, forced him to return home. Margaret and Norah Clegg, with their experience at the Hornsey housing trust, became leading figures in aiding refugees from northern Europe. Margaret told AV that she was almost living in the Town Hall and placing hundreds of men, women, and children in private homes as well as requisitioning houses and furnishing them.

Before he left Washington, Hill, strongly encouraged by Ambassador Lothian, was due to give the commencement speech at the California Institute of Technology. Hill regretted that his early departure meant that he could not deliver it in person on 7 June, but his words were read to the audience of 2,000 people. In his peroration, he cautioned them that the preservation of civilization entailed risk taking:

> Civilization will not perpetuate itself. It will continue only so long as its trustees are prepared at any time to make effort and sacrifice in its cause. There is no simple formula and "safety first," good though it be on the highways, is a disastrous guide to life.[49]

It went without saying that he was not going to shirk his own duty.

The first half of June saw the inexorable collapse of France. With Walter still in a camp about 300 miles south of Paris, Otto and Hedwig decided to head for Toulouse in the southwest. They were unable to board a train and instead hired a taxi driven by a young woman. Refugees carrying their meager possessions choked the roads so that the journey took two full days and nights. On arrival, Meyerhof estimated there were 100,000 displaced Belgians already in Toulouse. He and Hedwig rented one room with a single bed on a noisy street.[50] The choice of Toulouse, apart from putting distance between him and the invading German army, was because Professor Camille Soulla offered him a job in his physiology institute. Soulla was a good friend of Hill's and involved in the nascent French resistance movement.[51] Otto described the institute as modest and dirty, and although he was distracted by the events unfolding in northern France, he did attempt some mundane work in the laboratory and library. He decided to make a last-ditch effort to reach England. They traveled on an overcrowded train to Bordeaux, but there was no British steamer left in the port and the estuary was already mined. Next, they headed down to Bayonne which was over-flowing with desperate people. The British Counsel decreed that only British citizens would be taken on the last, overcrowded boat to leave. Otto broke down in tears.[52] They made their way back to Bordeaux, where there were rumors of an imminent German arrival, so they returned by car to Toulouse.

Otto wrote one dozen letters from Toulouse between 20 June and 10 July to colleagues in the United States, seeking an appointment that would bring an immigration visa. One was to the Rockefeller Foundation (RF):

> As I judge from the political situation, there will be in future no chance for me to keep my Paris position, even when I should escape from immediate danger. ... I would accept any position in the USA which would allow me to get a non-quota visa to live on a modest scale with my wife and my youngest son.[53]

Aside from the phrase "immediate danger," it is an unemotional letter, politely seeking an academic appointment through the RF. A sense of urgency was injected fortuitously by a telegram sent from London by Hill to Alan Gregg, the director of medicine at the RF:

> COULD YOU GET MEYERHOF, WIFE AND SON AWAY FROM TOULOUSE? IMPOSSIBLE FROM HERE[54]

Gregg sent both communications to A.N. Richards, the professor of pharmacology at the University of Pennsylvania, who had long established ties to the RF. Gregg stated that if Penn could offer Meyerhof a post (with visa), the RF would pay half his salary plus his travel costs. Richards received the letter on Saturday, 13 July, and after discussion with colleagues, he was able to confirm the appointment on the Monday. On 18 July, Thomas S. Gates, the president of the University of Pennsylvania, sent a cable and letter to Meyerhof confirming a research professorship with lifetime tenure and non-quota US visa. Gates advised him to make travel arrangements through Alexander Makinsky, the RF representative in Lisbon. Otto and Hedwig arrived in Marseille at the end of July, where they secured one of the few rooms available at the luxury Hotel Splendide, next to the train station. They would wait while Makinsky booked them a passage to America and planned to travel to Portugal as soon as it could be arranged. Otto wrote to Gottfried, telling him he had just received two tele-grams from Hill saying that he had been making vain attempts to get them to England; nevertheless, Otto believed there was a possibility of evacuation because "My friends are very active and had a very rapid success in their endeavour."[55] He went on to say, "Of course we must have courage and confidence into the future, and I hope, this will not lack for yourself." Hedwig added a short greeting, "My loved son, I am so sorry that I cannot be with you and care for you. But my love surrounds you."

In mid-June, Hill spent his first week back in England writing various memos regarding his recent assignment in North America. A few were novel, such as encouraging American volunteers to join the British services by dropping the requirement for them to pledge allegiance to the king. Pilots were especially needed and the presence of US flyers, he said, "would make America feel that the war was theirs—which is very desirable."[56] Hill had conferred with Justice Frankfurter on how to entice the United States into an exchange of technical information, and Hill made sure that his main reports were widely circulated in London and Washington. Now he did not feel constrained to write in bland prose; however:

> Our impudent assumption of superiority, and a failure to appreciate the easy terms on which closer American collaboration could be secured, may help to lose us the war. ... There is a very strong desire to help us, but of course the same fetish of secrecy exists in the Service departments in the USA as in Great Britain; it can be avoided in only one way, namely by a frank offer to exchange information and experience; without such an offer the resources of the US will remain imperfectly accessible to us.[57]

A copy reached Churchill, via Prof. Lindemann. He responded that "a decision should be postponed for a short period, and that the question as a whole should be treated as part of the larger issue [of American cooperation]." In the meantime, Hill pressed his proposals with the secretary of state for air, Sir Archibald Sinclair, who wrote to Churchill asking if the time was now "ripe." Churchill replied on 30 June that a technical mission could be sent to the United States, "provided the specific secrets and items of exchange are reported beforehand."[58]

In Washington, Lord Lothian lost no time in sending an *aide-memoire* to President Roosevelt indicating the British desire for a communication of secret technical information without any inherent bargaining. The choice to lead the British Technical Mission fell on Henry Tizard, even though he had recently resigned from the Air Ministry after sensing that he was being ignored and that Churchill's ubiquity was counterproductive. Hill showed his loyalty to Tizard and distaste for Lindemann by resigning from the now leaderless Tizard Committee on air warfare on 26 June.[59]

Lindemann was not displeased to see Tizard disappear from London. The mission included representatives of the three Services and scientists with knowledge of radar, most notably John Cockcroft. Tizard left England by flying boat for Newfoundland in mid-August and spent his first week in Ottawa. Ralph Fowler had arrived in Ottawa in July, as scientific attaché, bringing technical reports. After a few weeks' work, he wrote to Tizard to say that he had been ordered to bed for a month.[60] He recovered by early September and wrote to Hill from Ottawa saying: "Things are going half well, very well in fact, but the Americans have damned little to offer."[61] He was exasperated by their tendency to talk about projects still at the paper stage as though they were already in operation. Fowler despite his weak physical condition would play a vigorous part, often in company with Cockcroft, in discussing proximity fuzes, ordnance, submarine and aircraft detection, and, above all, applications of radar, up and down the eastern seaboard.[62] Cockcroft and Dr. E.G. Bowen (from Watson-Watt's group) sailed from Liverpool at the end of August to Halifax, bringing, literally, a black box that contained, inter alia, a prototype cavity magnetron that allowed the development of centimeter wavelength radar, vital to the Allied war effort. The Tizard Mission was a triumph in establishing technical cooperation between the British and Americans. Although Hill did not take a direct role, it is hard to imagine that it would have ever happened without his ceaseless promotion in the first half of 1940.

Since the summer of 1939, Sir William Bragg PRS, aided and sometimes led by his two secretaries, Egerton and Hill, had been trying to formalize an enhanced role for the Royal Society in liaising with government departments across a wide range of scientific issues (Figure 10.2).[63] Their efforts had met with some resistance from the secretaries of permanent government departments such as the Medical Research Council and the Department of Scientific and Industrial Research. As First Lord of the Admiralty, Churchill suggested at first the idea should be dropped, but Tizard managed to persuade him that there needed to be more coordination of research in the defense services, and he suggested Lord Hankey should be asked to oversee this. When he became prime minister, Churchill made his predecessor, Neville Chamberlain,

Figure 10.2 Officers of the Royal Society, 1939. (left to right) A.V. Hill (biological secretary), T.R. Merton (treasurer), William H. Bragg (president), A.C. Seward (foreign secretary), and A.C.G. "Jack" Egerton (physical secretary)
Photograph courtesy of the Wellcome Collection, London.

Lord President of Council (the minister responsible, among other duties, for government science). Chamberlain lost no time in reminding a hard-pressed Churchill about his own idea of appointing Lord Hankey. Hankey informed Bragg of the development, and Bragg wrote to Churchill on 10 June with a revised proposal, designed to be more palatable to the existing government research departments. He suggested a small, central, coordinating committee, headed by a cabinet minister, and comprising the three top officers from the Royal Society as well as the secretaries of the Agricultural Research Council, the DSIR, and the MRC.

After resigning from the committee on air defense, Hill mentioned that he might at some point raise the shoddy treatment of Tizard in the House of Commons.[64] The veiled threat led to a conversation with Air Minister Sinclair. Hill explained that it was not just a matter of Tizard, "but of the effective use of our scientific and technical assets and resources in waging war—and not squandering them in wild goose chases."[65] Sinclair offered to arrange for a small group of scientists to meet Churchill, but Hill rejected the proposal on the grounds that nothing would be achieved. Having told Sinclair that he was concerned by matters far more serious than personal disagreements between Tizard, on one side, and Churchill and Lindemann on the other, Hill proceeded to launch a vitriolic attack on the Prof. It came in the form of a memorandum, *On the Making of Technical Decisions by H.M. Government*, circulated to trusted friends and colleagues.

> It is unfortunate that Professor Lindemann, whose advice appears to be taken by the Cabinet in such matters, is completely out of touch with his scientific colleagues. He does not consult with them, he refuses to co-operate or to discuss matters with them, and it is the considered opinion, based on long experience, of a number of the most responsible and experienced among them that his judgment is too often unsound. They feel indeed that his methods and his influence are dangerous. ... He is gifted in explaining to the non-technical person; the expert, however, realises that his judgment is often gravely at fault. ... I realise that Professor Lindemann's presence may be indispensable to the Prime Minister, and the prestige and influence of the Prime Minister are now so important to the nation that some compromise may be necessary. ... I realise that the whole trouble is not due to one person, but that a system has grown up of taking sudden technical decisions of high importance without, or against, technical advice.[66]

Two appendices, laden with specific criticisms of Lindemann's ideas, were attached. He broadened his campaign against the general incomprehension of scientific advice in policy-making circles in another memorandum at the end of August.

> Practically none of the political leaders of the country have any personal acquaintance with science or technology, and their permanent scientific and technical staffs must either a) have it all their own way, or b) be merely the servants of the administration. Both alternatives are bad.[67]

This was more widely circulated and attracted the interest of the new secretary of state for air, Archibald Sinclair, who asked to meet Hill. In the meantime, Bragg's proposal for a small coordinating committee seemed to have withered on the vine. In September, Hill attempted to revive it by drafting his own proposals on "Scientific Research and Technical Developments in Government Departments."[68] On 21 September, he read a parliamentary answer given by Clement Attlee MP,[69] on Churchill's behalf, stating that proposals received from the Royal Society were under "active consideration." Hill immediately sent a handwritten note to Attlee on House of Commons notepaper,[70] questioning what was meant by "active consideration," as he was aware of Bragg's letter written to the prime minister in June, since when no response had appeared. For good measure, he included a copy of his latest memorandum on Scientific Advisory Councils. He also sent a copy to Neville Chamberlain, requesting a meeting, and stating he would be failing in his duty if he did not press for "a more confident use of our scientific resources."[71] Both Attlee and Chamberlain replied in supportive tones, with the latter claiming, "No one is more anxious than I am of the importance of harnessing Science to our chariot and making her one of the team that will win this war."[72] He added, in confidence, that he had just made a recommendation to Churchill which he hoped would be accepted and which would go a long way toward meeting Hill's views.

The general opinion in Whitehall was that another committee of scientists, however eminent, was probably unnecessary but that in order to placate them, it was worthwhile to create one. Even Lord Hankey, who was asked to chair the new committee, took this view and echoed the thought of the senior civil servant who wrote to him saying, "the government will have to set up some committee of the kind, if only to keep the scientific people quiet."[73] This was certainly true for A. V. Hill, who warned: "I propose to make myself a nuisance until something is done or I am squashed."[74] At the end of September, Churchill, who had many more pressing problems on his hands, acquiesced, but cautioned Chamberlain in a minute, "we are to have an additional support from the outside, rather than an incursion into our interior."[75] In the opinion of one historian, Hill's persistence, especially, ensured "a new era was begun in the relationships of science to government."[76] Chamberlain announced the formation of the Science Advisory Council in October with the officers of the Royal Society and presidents of the three government research councils as members. Hill believed it did make worthwhile contributions to the war effort but reflected that "it could have been much more useful if a scientific courtier had not monopolized the grace and favour of the Prime Minister."[77]

From 1933, Otto Meyerhof had waged his own bureaucratic struggle against the NSDAP in Heidelberg. In Marseille, he became trapped in a labyrinth of regulations with no obvious escape route from the Nazi Minotaur waiting to devour his family. The crux of the matter was Article 19 of the armistice signed by France on 22 June that became known as the "Surrender on Demand" clause. The rump of the French government, based in Vichy, agreed to hand over any German individual in France, identified by the Nazis. Meyerhof discovered this issue within days of arriving in

Marseille. He explained to Makinsky, the RF officer making his travel arrangements, "We are anxious to leave for Lisbon as soon as possible. On the other hand, it is quite difficult for foreigners residing in France to secure their exit permits at present. In my particular case, the Ministry of Education is trying to get the necessary authorization from the French Cabinet."[78] Makinsky had already encountered difficulties booking three passages across the Atlantic without having the necessary funds to hand. As he told his boss in New York there were about 20,000 people in Lisbon, all ready to sail to America. Many of them were wealthy and prepared to pay much more than the face value of a ticket, putting those who could only pay the official price at a disadvantage and making it impossible for those who could not pay full price at the time of booking.[79]

Despite appeals from the Rockefeller Foundation, the University of Pennsylvania, and the American ambassador to France, William C. Bullitt, the Vichy government refused to comply with the request for exit permits.[80] In his letter to Makinsky, Otto said that his son, Walter, was expected in Marseille in a few days and the plan was to organize his travel documents and passport too. Walter arrived by train, having just walked out of the internment camp at Le Cheylard. His German passport had been marked with a large J for Jew, and he had thrown it away a few weeks earlier. Now his father told him he had to return to the camp until a forged French ID document could be made for him.[81] Depressed, he left his parents for another few weeks until they signaled by postcard that he could return to Marseille.

The city was swollen with tens of thousands of desperate people, who had little chance of escape and who faced the prospect of being handed over to the Nazi occupiers or being sent to prison camps being opened around Vichy France. The Germans patrolled the waters around Marseille so that shipping ceased almost completely. General Franco's Spain was teetering on the verge of joining the Axis side. Portugal had concluded a non-aggression pact with Spain and was determined to remain neutral. Lisbon was the only remaining port in Western Europe that offered any chance of sailing to the United States. In order to reach Lisbon from France, one was supposed to have a valid passport, a safe-conduct pass and an exit visa from France, Spanish and Portuguese transit visas, and a visa to enter the United States. Most of those documents had to be renewed every three months at government offices that were rarely sympathetic to the plight of the would-be migrant. Otto and Hedwig were in possession of a non-quota US visa. Walter had none of the required documents, and the United States was no longer issuing visas on a quota basis.

Salvation came to the Meyerhof family, as it did to hundreds of Jewish refugees in France, through the politically committed, Harvard-educated, American journalist Varian Fry.[82] Dressed immaculately in a pinstriped suit that reflected his preppy background, Fry came to Marseille from New York with a list of 200 Jewish artists and intellectuals that the Emergency Rescue Committee he had helped to organize wanted to save. He intended to stay at the Hotel Splendide but had to wait until the end of August before a room became available. Rumors swept through the refugee community every day in Marseille, and none was more powerful than the apparition of an American carrying money and visas. Otto Meyerhof did not have to join the incessant stream flowing into the lobby inquiring for the American in Room 307; he just happened to meet him around the hotel. Fry informed Otto that he was not on his list, but he would help—saving a Jewish Nobel laureate would do no harm to fundraising in New York.

During his first week in Marseille, Fry was given a tutorial by another American volunteer, Frank Bohn, on the extensive, almost insurmountable, paperwork requirements faced by refugees. Bohn explained that while French exit visas were virtually impossible, the Spanish and Portuguese visas were more obtainable. Meyerhof also met two Unitarian ministers from Boston, Charles Joy and Waitstill Sharp, who ran the Unitarian Service Committee activities in Marseille and Lisbon. They persuaded Meyerhof that lack of a French exit visa should not deter him from trying to leave over the Pyrenees.[83] Fry seems to have conveyed this view to Meyerhof by early September because Otto told Dr. Joseph Stokes, a Philadelphia pediatrician passing through Marseille, that he had given up "all hope of ever securing a French exit visa."[84] Stokes also said that Meyerhof would be leaving Marseille around 7 September for the small fishing village of Banyuls-sur-Mer, where his address would be Laboratoire Arago, Biologie Marine. Otto had managed to arrange an appointment at the marine laboratory, which is on the rocky coastline just a few miles above the Spanish border.

A few days after they arrived, a young German couple appeared at the front door of the Meyerhofs' apartment. They introduced themselves as Hans and Lisa Fittko and said that Varian Fry had given them the address.[85] Lisa was Jewish, and Hans had been an active anti-fascist since 1933. They had been on the run across Europe for seven years and were now intending to cross into Spain. It was agreed that eighteen-year-old Walter would accompany them, and they left at noon the next day, posing as hikers. After several arduous hours, they were arrested by gendarmes, who asked what they were doing so near the border. The police proved unsympathetic when they explained they were Germans desperate to escape to Spain; they were taken to the village of Cerbère in the foothills of the Pyrenees. By sheer luck, while eating lunch with his captors at a café, Walter recognized a local customs officer who was a contact of Varian Fry's. He wrote a note on toilet paper he kept in his back pocket and dropped it on the ground. Fortunately, the customs officer picked it up and took it to his parents so that they knew Walter was going to be arraigned in Perpignan the next day. The Meyerhofs immediately went to see the mayor of Banyuls-sur-Mer, Vincent Azéma, who phoned the judge in Perpignan, a friend of his from the Socialist party. After the police explained that Walter and the Fittkos had been found trying to leave France illegally without exit visas, the judge thanked them and turned to the accused saying, "You are free to go."[86]

When Otto and Hedwig decided to make the walk across the Pyrenees in early October, Fry chose another resourceful guide for them. Léon Ball was an American, who had lived and worked in France for many years. Ball collected the Meyerhofs and two or three others from a hotel in Cerbère at 3 A.M. They were allowed one small handbag, told to walk single-file and not to talk.[87] As they approached the border after descending from the arduous trek over the mountains, Ball heard voices and made them hide in bushes before continuing barefoot. He turned back as they approached the Spanish border post at Portbou on the Costa Brava, early on a Sunday morning. The Spanish border officers were "extremely arrogant" and seemed disposed not to allow anyone to pass. While Otto was trying to make his case, Hedwig noticed a car drive up and approached the driver in tears. He was John Hurley, the US Consul General from Marseille,[88] making his own way to Portugal. Hedwig explained that her husband was a Nobel prizewinner who had accepted a professorship at the University of Pennsylvania, and that they were trying to reach Lisbon, where the Rockefeller Foundation was making travel arrangements to the United States. Hurley, a tough soldier from World War I, was not going to be overawed by Spanish border officials. The Meyerhofs were the only people from Ball's small band to be allowed across.

Notes

1. Hofmann et al. (2012).
2. M. Morange, "L'Institut de biologie physico-chimique de sa fondation à l'entrée dans l'ère moléculaire," *La revue pour l'histoire du CNRS* [online], 7/1/2002 (posted 17/10/2006, accessed 1/5/2019).
3. W. Meyerhof (2002), 16.
4. O.F. Meyerhof to O. Warburg (5/6/39) in Werner (1996), 326–328.
5. Deichmann (2002).
6. Hofmann et al. (2012).
7. O.F. Meyerhof to G. Meyerhof (3/9/39) MPT50 M613 U Penn archives.
8. W. Meyerhof (2002), 41.
9. O.F. Meyerhof to G. Meyerhof (11/9/39) MPT50 M613 U Penn archives.
10. O.F. Meyerhof to G. Meyerhof (11/9/39) MPT50 M613 U Penn archives.
11. O.F. Meyerhof to G. Meyerhof (17/10/39) MPT50 M613 U Penn archives.
12. O.F. Meyerhof to G. Meyerhof (15/2/40) MPT50 M613 U Penn archives.
13. O.F. Meyerhof to A.V. Hill (9/5/40) AVHL I 3/58.
14. O.F. Meyerhof to G. Meyerhof (14/5/40) MPT50 M613 U Penn archives.
15. W. Meyerhof (2002), 45.
16. O.F. Meyerhof to G. Meyerhof (17/5/40) MPT50 M613 U Penn archives.
17. Hofmann et al. (2012).
18. O.F. Meyerhof to G. Meyerhof (25/5/40) MPT50 M613 U Penn archives.
19. en.wikipedia.org/wiki/Cambridge_University_(UK_Parliament_constituency) (accessed 21/7/20).
20. A.V. Hill to Sir Geoffrey Ellis (16/1/40) AVHL II 5/47.
21. Sir Geoffrey Ellis to A.V. Hill (19/1/40) AVHL II 5/47.
22. A.V. Hill Draft speech, undated AVHL II 5/47.
23. *Cambridge Review* supplement (16/2/40) AVHL II 5/47.
24. D. Zimmerman, *Top Secret Exchange: The Tizard Mission and the Scientific War* (Montreal: McGill-Queens University Press, 1996).

25. A.V. Hill, "Science in the War," *Cambridge Review* (May 1941), reproduced in Hill (1960), 274–279.

26. Clark (1965), 250.

27. R.H. Fowler letter to A.V. Hill (22/2/40) AVHL I 2/7.

28. Zimmerman (1996), 53–54.

29. A.V. Hill to his family (26/3/40) AVHL II 4/2 CAC.

30. A.V. Hill to H.T. Tizard (28/3/40) Tizard personal papers AIR 20/2361 National Archives (NA).

31. Hill, "Science in the War," op. cit.

32. Zimmerman (1996), 55.

33. A.V. Hill to H.T. Tizard (23/4/40) Tizard personal papers AIR 20/2361 NA.

34. www.schenectadyhistory.org/military/worldwar2/esaw (accessed 27/7/20) .

35. A.V. Hill to his family (25/4/40) AVHL II 4/2.

36. Zimmerman (1996), 57–59.

37. H.T. Tizard to Secretary of State for Air (30/4/40) HTT Papers HTT 251 IWM.

38. H.T. Tizard to A.V. Hill (7/5/40) Tizard personal papers AIR 20/2361 NA.

39. A.V. Hill to his family (10/5/40) AVHL II 4/2.

40. A.V. Hill memo "Uranium–235" (16/5/40) AB 1/9 NA. Two weeks earlier, Vannevar Bush writing to a colleague stated his personal preference was "to do nothing" about the military value of fission, but thought the United States had to stay active in order "to know what others are doing. The whole thing may, may of course, fizzle." G. Pascal Zachary, *Endless Frontier: Vannevar Bush, Engineer of the American Century* (New York: Free Press, 1997), 203–204.

41. A.V. Hill to H.T. Tizard (16/5/40) TPP AIR 20/2361 NA.

42. R.H. Fowler to P.A.M. Dirac (25/1/1939) DRAC 3/7. In 1921, Fowler married Rutherford's only daughter, Eileen. During the 1920s, he became one of the country's leading theoretical physicists. He patiently guided the unworldly Paul Dirac, as he started in the field of quantum mechanics. Eileen died in the postpartum period after the birth of their fourth child at Christmastime, 1930. AV made sure that Fowler and his oldest child, Peter, came to stay at Three Corners in the summer of 1931.

43. A.V. Hill "R.H. Fowler (1889–1944)," in Hill (1960), 179.

44. A.V. Hill to H.T. Tizard (20/5/40); Clark (1965), 253.

45. H.T. Tizard to A.V. Hill (31/5/40) TPP AIR 20/2361 NA.

46. Zimmerman (1996), 61–70.

47. W.S. Churchill to A.V. Alexander (21/5/40) quoted in Zimmerman (1996), 64.

48. W.S. Churchill to W.L. Mackenzie King (5/6/40) quoted in Zimmerman (1996), 67.

49. A.V. Hill "The World of Tomorrow" (1940) M&R, 512–581.

50. O.F. Meyerhof to G. Meyerhof (9/6/40) MPT50 M613 U Penn archives.

51. Camille Soulla (1888–1963) M&R, 130–137.

52. Otto gave details of their efforts to go to England in a later letter to Gottfried (17/5/41). Walter in his 2002 memoir, *In the Shadow of Love*, gives a slightly confused account of their movements.

53. O.F. Meyerhof to R.A. Lambert (5/7/40), rockfound.rockarch.org/digital-library-listing/-/asset_publisher/yYxpQfeI4W8N/content/letter-from-otto-meyerhoff-to-robert-a-lambert-1940-july-05 (accessed 22/8/20).

54. "Speech Recounting Otto Meyerhof's Escape from Germany, Read by A.N. Richards," *100 Years: The Rockefeller Foundation*, rockfound.rockarch.org/digital-library-listing/- (accessed 25/6/2020).

55. O.F. Meyerhof to G. Meyerhof (27/7/40) MPT50 M613 U Penn archives.

56. A.V. Hill (18/6/40) AVHL I 2/1.

57. Clark (1965), 254–255.

58. Zimmerman (1996), 69–71.

59. Clark (1965), 241.

60. R.H. Fowler to H.T. Tizard (29/7/40) TPP AIR 20/2361 NA.

61. R.H. Fowler to A.V. Hill (11/9/40) AVHL I 3/19. AV replied to his friend on 4/10/40 that "I have had the warmest accounts both from Ottawa and from Washington of what you are doing with the National Research Council. I never doubted that you would, but it is always nice to find even one's certainties confirmed."

62. HTT 254 IWM.

63. W. McGucken, "The Royal Society and the Genesis of the Scientific Advisory Council to the War Cabinet, 1939–1940," *Notes and Records of the Royal Society of London* 33, no. 1 (1978): 87–115.

64. In 1941, when Tizard was experiencing an impossible time as the unpaid adviser to the Minister for Aircraft Production, Lord Beaverbrook, Hill praised him in a parliamentary debate on "Necessity for Research." "Tizard," he said, "found all sorts of difficulties in his way. He faced those difficulties with much more patience than I should have faced them. . . . The matter could be put right now if Sir Henry Tizard could be induced to return by offering him proper facilities and proper authority in his work. He is not like Achilles, sulking in his tent. He is unique, alike for his operational knowledge, for his knowledge of aeronautics and aerodynamics, for his technical and scientific knowledge, and for the complete confidence which his scientific colleagues and, I may say, the officers of the Royal Air Force, have in him. I submit that it is intolerable that we should not be using a man of his quality to the full, and that it is disgraceful that he should be driven from his position in the Ministry." api.parliament.uk/historic-hansard/commons/1941/mar/11/necessity-for-research.

65. Quoted in Clark (1965), 241.

66. Quoted in Clark (1965), 144.

67. A.V. Hill, "Scientific Advisory Councils in the Service and Supply Ministries" (1st draft 22/8/40) AVHL 2/1.

68. A.V. Hill, "Scientific Research and Technical Developments in Government Departments" (undated) AVHL I 2/1.

69. Clement Attlee (1883–1967), Leader of the Labour Party (1933–1955). He joined the War Cabinet as soon as Churchill became prime minister in May 1940 and was made deputy prime minister in the cabinet reshuffle of 1942. In 1945 he became prime minister after the Labour Party defeated Churchill's Conservatives in the "Khaki" General Election.

70. A.V. Hill to C R Attlee (21/9/40) AVHL I 2/1.

71. A.V. Hill to N. Chamberlain (21/9/40) CAB 21/829.

72. McGucken (1978).

73. Lord Hankey to Sir Alan Barlow (20/9/40) CAB 21/829 NA. Barlow and his wife Norah (daughter of Horace Darwin) were good friends of the Hills.

74. A.V. Hill to W/Cdr Elliot (10/9/40) CAB 21/829 NA.

75. W.S. Churchill to N. Chamberlain (27/9/40) quoted in McGucken (1978).

76. McGucken (1978). Equally influential was the effort to enlist public support for an increased role for science by the publication of a Penguin book, *Science in War*, which appeared that summer. It was written, anonymously, by two dozen younger and more

radical scientists, organized by Solly Zuckerman and J.D. Bernal. The 140-page paperback sold 100,000 copies within weeks; see B. Donovan, *Zuckerman: Scientist Extraordinary* (Bristol: BioScientifica, 2006), 41–44.

77. A.V. Hill, "Jack Egerton as Secretary of the Royal Society, 1938–1948" M&R: 94–104 AVHL I 5/4.

78. O.F. Meyerhof to A. Makinsky (1/8/40), rockfound.rockarch.org/digital-library-listing/-/asset_publisher/yYxpQfeI4W8N/content/letter-from-otto-meyerhof-to-alexander-makinsky-1940-august-01.

79. A. Makinsky to A. Gregg (31/7/40), rockfound.rockarch.org/digital-library-listing/-/asset_publisher/yYxpQfeI4W8N/content/letter-from-alexander-makinsky-to-alan-gregg-1940-july-31.

80. "Speech Recounting Otto Meyerhof's Escape from Germany." It is believed that not only did the Vichy government stop issuing exit visas to refugees from central Europe, it also alerted the armistice commission at Wiesbaden as to the whereabouts of applicants so that the Gestapo could pounce on them. www.yadvashem.org/download/education/conf/BrownVFry.pdf (accessed 25/8/20).

81. W. Meyerhof (2002), 47–50.

82. Varian Fry (1907–1967) Although his name is not as well known as it deserves to be, there is a wealth of biographical material available. I have relied mainly on two monographs: Sheila Issenberg, *A Hero of Our Own: The Story of Varian Fry* (New York: Random House, 2001), and Rosemary Sullivan, *Villa Air-Bel: World War II, Escape and a House in Marseille* (New York: HarperCollins, 2006)

83. R.A. Lambert, "Memorandum of interview with Otto Meyerhof (26/10/40)," rockfound.rockarch.org/digital-library-listing/-/asset_publisher/yYxpQfeI4W8N/content/memorandum-of-interview-with-otto-meyerhof (accessed 25/8/20).

84. A. Makinsky to A. Gregg (9/9/40).

85. W. Meyerhof (2002), 53–56.

86. After this first aborted escape (which Lisa Fittko overlooked in her 1985 memoir), she contacted Mayor Azéma who drew a map of an escape route across the Pyrenees for her. He also provided her and Hans an apartment over the Customs Office in Banyuls. Over six months, they guided hundreds of émigrés, referred by Fry, to safety, making up to three excursions per week (q.v. Sullivan, 2006).

87. rockfound.rockarch.org/digital-library-listing/- (see ref 73); W. Meyerhof (2002) and O. Meyerhof to G. Meyerhof (17/5/41) MPT50 M613 U Penn archives.

88. John P. Hurley (1878–1944) was forced out of Marseille in August 1940. Hiram "Harry" Bingham IV (1903–1988) was the deputy consul for visas, whom Varian Fry described as his "partner in crime" for supplying thousands of false passports to refugees, against State Department rules. See Isenberg (2001), 109; and photos.state.gov/libraries/estonia/99874/History%20stories/John-Patrick-Hurley.pdf and www.holocaustrescue.org/historic-background-of-rescue-in-france (accessed 26/8/20).

11
Deliverance

Otto and Hedwig took a train to Barcelona and, after a few days to rest, flew to Lisbon. They were met by Alexander Makinsky from the Rockefeller Foundation (RF), who managed to book them a passage on SS *Exochorda*, an American cargo-passenger ship due to sail to New York on 16 October. While staying at the Hotel Tivoli, Otto wrote to his sister Therese, who lived in London, with the story of their escape. He told her, "We are very sorry and dispirited that we had to leave Walter but hoping US friends will get him out soon."[1] Meyerhof wrote a note to Makinsky,[2] just before the ship left Lisbon, thanking him for the "incessant help and interest" he had displayed toward Hedwig and himself. The trans-Atlantic voyage took ten days and was rough throughout, but the Meyerhofs' spirits were lifted immeasurably by a reception party waiting for them when the *Exochorda* docked in Jersey City. It was led by their daughter, Bettina, along with David Nachmansohn and Fritz Lipmann, another associate from his time in Dahlem, plus Dean Burk from the NCI, and representatives of the RF.

That evening, Otto was debriefed by Robert A. Lambert, the deputy director of the RF's medical sciences division. He was seeking detailed information about Meyerhof's escape that might aid other scholars. Lambert formed the opinion that while the French might not issue an exit visa, there were individuals such as Meyerhof they would be relieved to see the back of, and the Spanish were more likely to admit those who already possessed a US immigration visa. But the journey was becoming more difficult and "the barrier may become insurmountable at any time." He asked Meyerhof about the Quakers aid group in Marseille. He replied they were not helpful because they insisted on transparency and honesty—unlike the Unitarians who did "not look upon a subterfuge under present conditions as an unpardonable sin." Meyerhof was undecided on the question of encouraging French scientists to leave but urged the RF to continue supporting their laboratories with apparatus and funding. On reflection, he favored inviting first "those whose known political views might endanger their freedom to work—or even their lives."[3]

Within one month of arriving in Philadelphia, Meyerhof was busy in his lab and giving talks. At the end of November, he gave a lecture to 200 biochemists and physiologists (including one dozen former students) at Cornell Medical School and also spoke at Yale and Swarthmore. He lost no opportunity to praise the British, now standing alone against Germany. He wrote to Gottfried, "we must be confident to the final outcome and the composed attitude of the English people and the aliens-friends, suffering with them is the best guarantor thereof. A.V. Hill was much pleased

Bound by Muscle. Andrew Brown, Oxford University Press. © Oxford University Press 2023.
DOI: 10.1093/oso/9780197582633.003.0011

that I am here always preaching more help for England, indeed the too-short, weekly working time of 40 hours is the greatest handicap for this."[4]

His overriding concern, however, was to rescue Walter from Vichy France. Walter was in touch with Charles Joy at the Marseille office of the Unitarian Service Committee, who summed up his predicament by saying that Walter had no French ID card and no other papers of any sort.[5] He took him to the American consulate, where there seemed to be a possibility of a US quota visa, but Joy thought obtaining a Spanish immigration visa looked "insuperable." Walter returned to Banyuls to await events and was advised to return to the consulate in December. He was then granted a visa, valid until the beginning of May 1941, because US Secretary of State Cordell Hull sent a telegram instructing the staff to issue one. Walter wrote to his sister, Bettina, at Johns Hopkins that "Father's ghost is opening all doors for me." He wrote to Varian Fry at the end of January to say that he had just secured a French exit visa to go with his American and Portuguese visas, but his application for a Spanish transit visa had been refused. He also informed his father, who attempted to bring diplomatic pressure to bear through the Spanish embassy in Washington and the US embassy in Madrid. These were unsuccessful and a second application was refused in mid-February. At that point, Walter left Banyuls-sur-Mer for Marseille and worked to help the Unitarian Service Committee. Fry constantly encouraged him that he would eventually get out and tried to ensure that he had food and places to stay. At times that would include visits to the Villa Air-Bel outside the city, where famous artists such as Max Ernst and Marc Chagall waited for Fry to organize their escapes. Walter, at the age of eighteen years, was too shy to mix in such outlandish company. In March, he received the Spanish visa valid for ten days. He took a train and then a bus to Le Perthus, a quiet border crossing, where after a wait of five hours while the officer checked with Madrid, he was refused entry because they did not recognize the "Affidavit in lieu of Passport" issued by the US Consulate.

A dejected Walter returned to Banyuls and cabled Joy telling him what happened. He responded immediately that he was to return to Lisbon from Marseille in two days and suggested Walter should accompany him as his secretary. At the station, Joy purchased two tickets to Barcelona and bribed the Thomas Cook agent with $40 "to see to it that that Spanish border police would not make any difficulties for his secretary." In Lisbon, Makinsky found a berth for Walter on a Portuguese cargo ship, two days before his US visa expired: he finally arrived in Philadelphia in mid-May. Otto wrote to Gottfried to tell him the news but expected that "England knows" because a cable had already been sent to Hill.[6] Walter moved in with his parents in Philadelphia and became a graduate student in physics at Penn.

When Hill arrived back in England on 13 June 1940, he found a country under imminent threat of invasion. Before the end of May, Churchill gave the order to "collar the lot," meaning the urgent internment of tens of thousands of Germans and Austrians living in the United Kingdom, the great majority of whom had previously been judged by efficiently run Aliens Tribunals to present no security threat: indeed, most were implacable opponents of the Nazis. Italians were included after Mussolini joined the Axis on 10 June.[7] Scooped up in the process were more than 500 scholars supported by SPSL, and they became Hill's major concern. Esther Simpson wrote to him from the SPSL office saying, "we have had a great need for you in the perplexities which the various internment of refugees and other measures have plunged us."[8] A few days later, Hill wrote to Sir Alexander Maxwell, permanent under-secretary at the Home Office,[9] suggesting that SPSL should prepare a dossier of their scholars, giving evidence on an individual basis of their personal integrity and the significance of their work. Maxwell thought the new restrictions were unjust and readily accepted Hill's offer.

A handful of MPs, led by Major Victor Cazalet (Conservative) and Eleanor Rathbone (Independent MP for Combined English Universities), grilled Sir John Anderson, the Home Secretary, about the changes in policy. A humane man, Anderson plainly understood its imperfections. The tragic sinking on 3 July of the SS *Arandora Star* by a U-boat off the northwest coast of Ireland, while she was carrying 1,500 enemy aliens to be interned in Canada, gave rise to wider misgivings. Hill made his first intervention in the House on 18 July, when he asked Anderson whether new consideration for the release of internees involved in work of national importance would extend to those making "significant contributions to science and learning."[10] Anderson, with a slight smile, answered in the affirmative, which Hill saw as the key to unlocking as many SPSL internees as possible. Heinz Schild, a demonstrator in the pharmacology department at UCL, had been interned on the Isle of Man. His wife, who had just had a baby, was living with family in Wales and was informed that Heinz drowned in the *Arandora Star*. Fortunately, she soon heard that he was still in the Isle of Man and wrote to Esther Simpson at SPSL. Simpson replied that "Professor Hill has already raised your husband's case specially with the Under Secretary of State. He has told him of his error in reporting your husband lost and urged him to release Dr Schild. … The whole machinery of release is very slow."[11] Hill wrote encouraging letters to Schild during his internment. When he was released in January 1941, after additional lobbying from Dale and the Committee of Vice-Chancellors, Hill wrote to Mrs. Schild saying how pleased UCL was to have him back teaching pharmacology.[12]

Esther Simpson was tireless in obtaining vouches of personal integrity and loyalty to Britain for hundreds of interned scholars from their colleagues, especially if they were associated with august institutions such as the British Academy or the Royal Society. At the Royal Society, Hill organized a committee to prepare reports for the Home Office on interned scientists.[13] He cautioned Simpson that the reputations of the Royal Society and SPSL were both at stake. The material they submitted to the Home Office had to be convincing: the Royal Society needed to know the details of a scientist's work and would not be prepared to merely issue opinions. Beveridge recalled, "the work of the SPSL in the summer and autumn of 1940, was concentrated on preparing cases for these tribunals and applying for release of those refugee scholars on our register whose cases were approved by the tribunals."[14] Simpson's unstinting efforts produced four large bound volumes of biographical material that eventually resulted in the clearance of every SPSL-sponsored individual. Nine copies of each application had to be submitted to the Home Office, together with supporting documents. Hill wrote to her in admiration in September, saying, "I marvel at what you get done."[15] Between June and December 1940, she wrote sixty-one letters to Hill, often concerned with individual cases, and he sent her nearly fifty letters in reply.[16]

Hill continued to become involved in individual cases over the next two years. Early in 1942, efforts were made to have Maurice Pirenne, a biophysicist, released from the Belgian military based in England so that he could become involved in research on night vision (on which he was a recognized expert) at Cambridge. During a four-month period, Hill wrote more than one dozen letters on his behalf to the Belgian authorities, the MRC, Cambridge University, the Foreign Office, and various individuals. Pirenne wrote to him to say "the thoroughness and tenacity of your efforts fills me with optimism and gratitude"[17] before Hill eventually prevailed.

Hill also raised his voice in support of the tens of thousands of interned aliens who had no connection with the SPSL. While accepting the need for widespread internment in May 1940, he resented the level of panic he identified as arising among the "ruling clique in Britain ... who were deceived by Hitler and Mussolini for so long." They were now misjudging "many of the bitterest enemies of the dictators ... whose one desire is to help in defeating our common enemies."[18] His weightiest intervention came in a House of Commons debate on 3 December 1940. Eschewing any definition of the war aims, Hill wanted to set out the principles on which the ultimate peace should be based, in order to encourage not just British citizens but people across the world. Failing to do so, he feared, would mean "our actions will be hesitant, timid and inconsistent. ... We have to set against Hitler's doctrine of racial superiority something better and more credible, and something which can move the hearts of common men."[19] He then highlighted the lack of a clear policy regarding alien internees as an example:

> Are we to accept their help or not? Is this fight their fight as well as ours or not ... Are we to give them the opportunity to serve the common cause or follow them about with petty restrictions and supervision as though they belonged to an inferior race?

He catalogued some examples of wasted manpower: for example, a physician who fled Nazism, obtained British qualifications, joined the Emergency Medical Service in 1939 (giving up his own practice), only to be interned for six months, before being told he could not rejoin the EMS. Through his work with SPSL and the Royal Society, Hill had become frustrated by the delays inherent in the appeals process. He told the House about an application from an internee on the Isle of Man that had taken forty-two days to arrive in London. He calculated that this "works out at a speed of a quarter of a mile per hour, or less than the speed of a tortoise." His witticism was challenged in the *Daily Telegraph*, citing the top speed achieved in a tortoise derby, suggesting, "Dr. Hill ought to withdraw his allegations and make it clear that Dr. Herbert Morrison's department moves three and half times faster than the speed of a tortoise."[20] As a physiologist who liked to think of himself as a humorist, AV could not let the opportunity pass and responded that his tortoise lived in Greece, where the hotter weather meant they moved faster.

The Blitz of London started in September 1940, and Margaret immediately became even more involved in housing the homeless. She wrote to her parents a few days later, life at Hurstbourne was "a bit hectic":[21] there were nine people sleeping in beds in the cellar; Polly (who was working at the Board of Trade) stayed in the wine cellar; her sister Janet (about to qualify as a physician) in the old pantry; Norah and Margaret slept in their office with two maids in the passageway; David in the drawing room; and AV slept in the front hall. Maurice, not mentioned in the letter, had not returned to Cambridge when war was declared, and instead volunteered for the Royal Navy. He was sharing an apartment with his cousin, Richard Keynes, in Weymouth. They were both employed at the Anti-Submarine Experimental Establishment on Portland Bill. David moved to the Postgraduate Hospital in Hammersmith in March 1940 to study the changes in circulating blood volume caused by crush injuries, which were now becoming commonplace. He contributed several significant research papers,[22] but his father was about to move him even closer to the sharp end of the war. In a letter to Gottfried, Otto mentioned AV was "proud how his whole family works for the country" before adding, "I am sure that he has a great share of merit for our now settling here."[23]

Hill was the chairman of the Ballistics Committee at the Ministry of Supply, which held a meeting on 7 August 1940. He invited General "Tim" Pile,[24] who oversaw A.A. defense for London, to attend because Pile had expressed concerns about the adequacy of his own service, given the recent destruction meted out by the Luftwaffe against European cities. At the meeting, Pile stated that the basis for A.A. fire was illusory since the Germans rarely followed a steady course and generally dive-bombed. He remarked sharply that: "Science should be capable of giving an A.A. officer a 10% chance of a hit."[25] Patrick Blackett, a committee member, had a short conversation with Pile, who told Hill after the meeting that Blackett was the obvious choice to be his science adviser. Hill agreed and wrote to Blackett saying, "Apparently it was a case of love at first sight."[26] There was a shortage of young physicists because so many were already involved in radar work, so Hill provided Blackett and A.A. Command with a trio of young, numerate physiologists: his son David, Leonard Bayliss, and Andrew Huxley. They became leaders in "Blackett's Circus," their main task being to improve the marriage of radar data derived from the attacking bombers with the gun laying controls of A.A. batteries. AV also maintained an interest in A.A. defenses, visiting the gun batteries around various ports in the southwest of England in August 1940—activities he combined with a short vacation at Three Corners. It would be the last visit he made there until after the war; the house was soon occupied by families who had been bombed out of their homes in Plymouth.

Figure 11.1 Observing a demonstration of A.A. fire. (left to right) Prof. Frederick Lindemann, Air Chief Marshal Sir Charles Portal, Admiral of the Fleet Sir Dudley Pound, Prime Minister Winston Churchill

Photograph by W.G. Horton (undated) courtesy of the Imperial War Museum, London.

The Director of Artillery, General Edward Clarke, wrote to Hill in October concerned about the futility of a new A.A. weapons system known as the Unrotated Projectile (UP). It was the latest iteration of Lindemann's pet idea of aerial mines, which in this case would be fired by rockets. The Prof had just organized a demonstration where the target conditions were ridiculously easy, and now "Winston has ordered a repeat trial for his personal attendance." (Figure 11.1)[27]

Clarke was worried that there was already a commitment "to a vast programme of UP equipment and to raising several thousand men in special units for manning it." This letter probably contributed to a memo of frustration about the realities of A.A. defense that Hill composed. After dismissing the notion that A.A. guns would ever destroy significant numbers of aircraft, their main purpose being as a hindrance or deterrent, Hill proceeded to demolish the notion of A.A. barrage. In his opinion, "the word ought to be dropped: it gives a false impression and is based on sloppy thinking and bad arithmetic."[28] Pointing out that the lethal zone of an exploding shell was only a few thousand cubic yards, lasting for a fraction of a second, he calculated to bring down a plane flying at 250 mph in a rectangle 10 × 4 miles high required about 3,000 shells per second to be fired to give a 1:50 chance of success. Hill conceded that a proximity fuse might increase the effectiveness, but that "proximity" to the target would still be required.

The disagreement over the worth of UP continued for months, and General Clarke wrote to Hill again in the summer of 1941. He mentioned that his new boss at the Ministry of Supply, Lord Beaverbrook, would welcome a discussion with Hill on the subject. Clarke was diplomatic but did disclose that he had told Beaverbrook "that the whole UP question had been distorted by backstairs politics and other influences and at times looked more like a 'racket' than a rocket."[29] The "Beaver" agreed with him and was flat against "fancy weapons." Beaverbrook phoned Lindemann to tell him so in the middle of a meeting and hung up before the Prof could reply. Hill wrote to Beaverbrook suggesting that an independent group should be set up to examine the question and take evidence from the General Staff. Beaverbrook declined the offer, saying that Harold Macmillan launched an inquiry one week previously into the use of weapons by the War Office and he did not want to compromise that.[30] UP was abandoned before the end of the year.

Hill often used to say that he became an MP to make a nuisance of himself. He certainly lived up to his billing as an Independent Conservative and was often a gadfly, tormenting the government over its prosecution of the war. His sallies would include personal attacks on Lindemann—something he engaged in far more than Tizard, whom Hill continued to believe was undervalued by Churchill. Hill rehearsed his ideas, perhaps to release his frustrations, in a series of undated memos. The first pair, "On the Making of Technical Decisions by H.M. Government" and "Scientific Advisory Councils in the Service and Supply Ministries," were circulated, as we have seen, to a small, influential circle in the summer of 1940. They were critical of Churchill's absolute reliance on Lindemann (elevated to the peerage as Lord Cherwell in 1941), and of the inadequate understanding of science among politicians in general.

Hill did not expound these radical ideas publicly in 1941. In his speeches and questions in the House, he concentrated on continuing questions about the under-employment of aliens and especially on public health. He made regular speeches supporting vaccinations for troops and children, blaming "anti-vivisectionists" for casting doubt. On occasion, as in his speech during an adjournment debate in May 1941, he managed to combine all these topics:

> We know that 3,000 deaths from diphtheria and 60,000 cases annually could be abolished if only we could think ahead, if the Government and the people were not so complacent about the situation and if doctors were available to carry out the necessary immunisation. But if we are to avoid disease, if we are

to treat the wounds of war and disease resulting from fatigue, from temporary food deficiencies, from abnormal conditions due to enemy action resulting, for example, in the cutting of water supplies and drainage, if we are to provide the Fighting Services, our ships, the Royal Air Force, the units of the Army at home, and more particularly in the tropics where there are new dangers to meet, then far more doctors are required. For this reason, we can only welcome with gratitude the promise of the United States to send us 1,000 doctors to help us. This promise shows the degree of realisation by the United States of our need. In view of this realisation it seems to me appalling that we should have 1,300 doctors from Europe unemployed of the 1,400 available. ... The old traditional government of the medical profession, or, as it should be termed, the calling of medicine, by wealthy consultants is already doomed; we must realise that the public health, and not the interests of consultants in Harley Street, or even the supposed interests of busy practitioners who are paid by the job and not by the day, are at stake.[31]

In September, Hill, as a member of the Scientific Advisory Committee (SAC), took part in seven meetings chaired by Lord Hankey to assess the MAUD report on the feasibility of making an atomic bomb. The report, drafted by James Chadwick who had overseen the British investigation of the practicability of nuclear fission from Liverpool University, concluded that a uranium bomb was realistic and likely to lead to decisive results in the war. Arguing that work on such a weapon should be expanded and given the highest priority, the report also recommended increased collaboration with America.[32] SAC questioned several witnesses on the unresolved technical issues, including Chadwick, Fowler, and Blackett. It also raised the issue of whether the large industrial plant necessary to separate ^{235}U, the fissile isotope present as less than 1 percent of natural uranium, could be built and effectively protected in wartime Britain. The weight of opinion was firmly against that idea, but one witness fervently in favor was Lord Cherwell, the Prof, who regarded American engineers as competent but was against reliance on another nation where such a powerful weapon was concerned.[33]

Hill's opinion is not recorded individually but it is easy to imagine, given his past efforts, that he would have strongly supported the recommendation in the final SAC report that there should be "close collaboration with the Canadian and United States Governments." He would have been furious but not surprised if he had known that the Prof had preempted the SAC review by minuting Churchill in August, as soon as he saw the MAUD report. Cherwell was convinced that the bomb could be built and was "strongly of the opinion that we should erect the plant in England or worst in Canada" despite the obvious wartime risks. Possession of such a weapon would mean the ability "to dictate terms to the rest of the world" so that "however much I may trust my neighbour, and depend on him, I am very much averse to putting myself completely at his mercy."[34] The careful conclusions of SAC, that there were still unresolved technical issues and that pilot plants for isotope separation should be built in Britain and Canada, before continuing with a full-scale plant in Canada, were ignored. Churchill had already forwarded Cherwell's minute to the Chiefs of Staff at the end of August with the comment: "Although personally I am quite content with the existing explosives, I feel I must not stand in the path of improvement, and I therefore think that action should be taken in the sense proposed by Lord Cherwell."[35]

Hill's friend from undergraduate days, Charles G. Darwin, replaced Fowler as the British scientific representative in North America, and together with Professor Oliphant, who was on a summer visit from Birmingham, they successfully lobbied leading scientific figures in the United States about the significance of the MAUD report.[36] Largely as a result of their interventions, President Roosevelt suggested to Churchill in October that efforts toward an atomic bomb "may be coordinated or even jointly conducted." His offer was essentially ignored in London because, as the official historian, Margaret Gowing, wrote later, "when the British nuclear scientists were in the lead ... the British failed to seize the opportunity."[37] This was precisely the kind of false sense of superiority that Hill warned Tizard about in 1940, when urging that a technological mission should be dispatched to America without preconditions. Ironically, there were again mutterings about security risks in the United States since it was still not a belligerent in October 1941, but Hankey's private secretary, John Cairncross, had already passed the MAUD report to Moscow.[38]

The attack on Pearl Harbor on 7 December brought the United States into the war, and while this was ultimately a decisive event, the fortunes of war did not turn quickly. Otto Meyerhof wrote to Gottfried from Philadelphia:

> This is indeed a grave low; but in spite of the new enemy in the East ... the outcome in total is considerably improved by the all-out war to which the USA is now forced by the Axis. Never would it have been possible to mobilize the resources of the USA to the last dollar and the last man, if this country had stayed on in peace and only supplied so called all-out-aid to Britain.[39]

Churchill paid a month-long visit to North America. He left London on 12 December, two days after the crushing news that Japanese aircraft had sunk two capital ships of the Royal Navy off the coast of Malaya; on his return, he learned that the Japanese Army was advancing overland toward Singapore—it would fall within one month. Churchill demanded a vote of confidence at the end of January because, as he told the House, "Things have gone badly and worse is to come."[40] Although there was only one vote against him in the confidence motion, he was obliged to make a major overhaul of his war cabinet. He brought Lord Beaverbrook back as Minister of War Production and streamlined his team.

While these events were unfolding, Hill was moved to compose a broad, fiery critique, "Combined Operations and a Great General Staff."[41] Befitting a one-time Third Wrangler, this began with an axiom: "*In modern war an accurate knowledge of weapons and equipment, their functions, their limitations and their availability, is fundamental to strategy and tactics."* He argued:

> Such knowledge is not possible in a single individual, particularly as a part-time job at the age of 67; it requires a full-time technical section of a combined General Staff, composed of young officers who have grown up with modern arms.

In case there was any doubt that Churchill was his target, he insisted that the prime minister's role as minister of defence *"should be abolished."* He charged the prime minister and Lord Cherwell together with being responsible for massive, wasted expenditure. "The Prime Minister's fertile brain is quite unsuited to well considered technical decisions on weapons," he wrote, "and Lord Cherwell's judgment on them has been shown by events to be disastrous, as his scientific colleagues foresaw." One year after the Battle of Britain, Hill thought *"all* operations now, whether by land or by sea, involve the use of the Air arm" and he saw this as a reason not to have an independent RAF. "The idea of bombing a well-defended enemy into submission, or even of doing serious damage to him, is complete illusion—and a very wasteful one." Nor was he any kinder about the senior service, saying too many admirals had been brought up in the old traditions so that they "persistently clung to the large capital ship as the basis of the fleet. Without strong air protection these precious ships are the gravest liability. The basis of the fleet of the future will be the aircraft carrier." The RAF and the Royal Navy did receive some credit for incorporating operational research unlike the Army (with the honorable exception of A.A. Command). There needed to be brighter minds with technical knowledge at the War Office, where the current secretary of state, David Margesson, who had been an appeaser before the war, was "completely incapable of understanding such matters." He wanted a Great General Staff to be appointed, emphasizing that *"its chief should be in the War Cabinet."*

He circulated his paper to J.M. Keynes, various scientific colleagues, and to several senior military officers. One was Air Chief Marshall Philip Joubert de la Ferté, who oversaw RAF Coastal Command, where he was determined to improve the ineffectual aerial attacks on U-boats in the Battle of the Atlantic. Joubert captured Patrick Blackett from A.A. Command in the spring of 1941 to serve as his personal scientific adviser.[42] The tactical improvements that Blackett and his team introduced, such as setting depth charges to explode at 25 rather than 100 feet underwater and making antisubmarine aircraft less visible in daylight by painting their undersides white, convinced Joubert of the value of operational research. He responded to Hill:

> What a firebrand you are—but I am delighted with your paper. Whether such a Grand General Staff as you envisage is a practical proposition must be a matter of opinion.... If you see me lunching with Lord Cherwell at the Athenaeum, please do not think I agree with his policies. ... I want to persuade him that the air/sea war is more important that the bombing offensive.[43]

"Combined Operations and a Great General Staff" formed the basis of a speech Hill made in parliament on 24 February 1942. The prime minister had opened the debate to explain his rationale for the recent cabinet changes. He then reported on his own roles, especially the unique combination of offices that he held. Seeking to clear up misunderstandings, Churchill explained:

> I may say, first of all, that there is nothing which I do or have done as Minister of Defence which I could not do as Prime Minister. As Prime Minister, I am able to deal easily and smoothly with the three Service Departments, without prejudice to the constitutional responsibilities of the Secretaries of State for War and Air and the First Lord of the Admiralty. I have not, therefore, found the need of defining formally or precisely the relationship between the office of Minister of Defence when held by a Prime Minister and the three Service Departments. I have not found it necessary to define this relationship as would be necessary in the case of any Minister of Defence who was not also Prime Minister. There is, of course, no Ministry of Defence, and the three Service Departments remain autonomous.[44]

Churchill maintained the present arrangement was "the best that can be devised to meet the extraordinary difficulties and dangers through which we are passing": there was absolutely no question of changing it while he remained in charge. He then moved on to review the general war situation, projecting "a sober and reasonable prospect of complete and final victory" after "a good many months of sorrow and suffering." In a crowded house, Hill sat through several hours of his fellow MPs, mostly veterans of the 1914–1918 war, commenting on the recent setbacks, the ministerial changes, how relations with new ally Russia were being developed with too much or not enough care, as well as the needs of the British people, before he managed to catch the Speaker's eye. He rose to welcome the announcement of a reshuffled War Cabinet but did not believe the changes were drastic enough. Before coming to the thrust of his paper, "Combined Operations and a Great General Staff," he made sarcastic reference to Churchill continuing as minister of defence, without naming him. Doubting that the use of "brave adjectives" impressed the enemy at all, Hill continued:

> Nor can the sort of technical knowledge which is necessary for those who have to guide our strategy now be acquired as a part-time job by an elder statesman whose historical outlook inevitably leads him to think in terms of earlier wars ... it is almost better to forget our history altogether than to act as though the strategy and tactics of the present war were similar to those of Agincourt, Waterloo or the Marne.

For Hill there was no romance in war, yet he underestimated the inspiring genius of Churchill's oratory.

Hill also suggested that the public should be told bluntly that the war would not be won by defensive measures. The Luftwaffe paid a heavy price for its daytime bombing raids in 1940, and the last nine months had seen little in the way of night raids. He pointed out that there were now hundreds of thousands of able-bodied men and women employed, or rather idle, waiting for further raids that seemed unlikely to occur. He made the case that if 20 percent of those in civil defense were transferred to producing weapons or making food, the population would only be marginally less safe. He did not shy away from bluntness himself: "We think 40,000 or 50,000 civilians killed far worse than 70,000 soldiers captured at Singapore, which, from the point of view of winning the war, is far more important."

Prior to his speech, Hill had condensed his central criticisms in a letter to *The Times*, cosigned by Sir Ralph Glyn, another Conservative MP who had lost patience with the Churchill's approach to the war. They claimed:

> True cooperation between Navy, Army and Air Force cannot be effective so long as each Service remains separate up to the level of the Chiefs of Staff Committee. ... The Prime Minister has decided to continue as Minister of Defence but without a trained and expert staff in the Ministry of Defence, his burden of responsibilities will continue to be insupportable and the conduct of the war will continue to suffer. The new War Cabinet is a great improvement on the old, but it still needs the advice and help of practical men working in combination and representing the most modern outlook, particularly in the technical sense, of the three Fighting Services.[45]

This letter together with the speech in parliament provoked second thoughts from Air Marshal Joubert, who wrote to AV that he believed "firmly that the Prime Minister must, in the last analysis, be responsible for defence matters." He also warned:

> We have arrived at a deplorable situation where, as a result of the criticisms made by ill-informed persons, the junior people in the Services are beginning to mistrust their seniors and the public at large to distrust the services generally. I do look to people like you, able, intelligent and thoughtful and with sufficient influence to mold public opinion in the right shape, to put this situation right.[46]

One of the most surprising changes in the shake-up of the War Cabinet was not that Churchill decided to sack David Margesson as secretary of state for war, but that he chose an under-secretary at the War Office, Sir James Grigg, to replace him. Grigg was a clever, career civil servant (like Hill, a top Wrangler at Cambridge in his youth), who had been Churchill's private secretary in the 1920s. He was quickly installed as an MP for Cardiff so that he could take up his new position. Hill sent him a memo setting out what he saw as the main principles for operational research that should be adopted at the War Office.[47] Grigg responded that he was in general agreement with Hill but would not join "the crusade for the revival of the M.G.O."[48] The master of the general ordnance was very senior army position first introduced by Henry V and abolished by Secretary of State for War Hore-Belisha in 1939. It is ironic that Hill was pushing for the resurrection of a fifteenth-century post, given his sly remarks about Agincourt and history in the House. Undaunted, Hill informed Grigg, "The General Staff of the War Office are called upon to express military requirements in technical terms, a function which their training has not fitted them to perform."[49] He observed that the few army officers with technical qualifications were either misplaced at the Ministry of Supply or wasted on routine maintenance work. He included a recent report analyzing technical errors in the Army that had resulted in an inferior Sten gun, and poorly designed valve and gear motor components.

Hill had been circulating his memos to an increasingly wide audience that included Robert Barrington-Ward, the editor of *The Times*. He requested that Hill should compress his ideas into a letter, summarizing his main recommendations.[50] That letter appeared on 21 April and argued, "what is in fact wanted is an integrated War Staff, a High Command, designed for the single purpose of planning the war as a whole, not further improvisation with existing machinery." New posts would include a chief of technical staff and a director general of OR, among others. Air Marshal Joubert cautioned AV again after reading it: "I am afraid, and this is entirely for your own personal consumption, that if you start tilting directly at Lindemann we shall not get anywhere."[51]

Hill's idea for a Great General Staff with executive powers to direct war strategy, rather than the existing arrangement of the service staffs briefing their chiefs of staff independently, was aired extensively in a House of Lords debate on 5 May. The debate on a government White Paper "Organization for Joint Planning"[52] was opened by Lord Denham, who soon cited Hill:

> There was a letter in *The Times* on April 21, from Professor A.V. Hill, which I will venture to quote to your Lordships as it bears on this topic. Professor Hill expressed the view that the use of the term "Combined General Staff" showed that the nature of the need was not fully realized. What, he said, is wanted is "an integrated War Staff ... a high command, designed for the purpose of planning the war as a whole, not further additions to existing machinery."

Denham was critical that the White Paper overlooked the need for scientific and technical input, a point also raised by the next speaker Viscount Swinton, who of course knew Hill well from the Tizard Committee. Swinton thought that the scientists were not being allowed to play their full part, or "at any rate, they do not think that they are." Swinton captured the general feeling that there should be a full-time chairman of the General Staff to assist the military chiefs, but that the prime minister and War Cabinet would remain supreme. It was a question of organizing the best machinery for allowing the best men to function optimally, rather than replacing individuals.

Viscount Trenchard, often a cantankerous man, sometimes referred to as "the father of the RAF" and its sole representative in the House of Lords, also spoke in the debate and was broadly in agreement with Denham and Swinton. He and Hill were friends, although Hill did not agree with him that the RAF bombing campaign was going to lead to the defeat of Germany. Trenchard had tried to dissuade Hill from taking part in the House of Commons debate at the end of February, especially "if you talk against bombing because I think I could tell you some facts that would make you think. I am perfectly sure on that one point you are absolutely wrong from the figures and facts I have got."[53]

Churchill took the decision to embark on the heavy bombing of German cities in February 1942; the policy was enthusiastically endorsed by Cherwell in a memo predicting heavy bombing of the largest fifty-eight German towns would render about one-third of the total population homeless. In the Prof's opinion, "this would break the spirit of the people."[54] Blackett and Tizard were both given the opportunity to review Cherwell's memorandum and found his figures for dehousing to be 5–6 times too high. Blackett was convinced that the "very ineffective offensive against enemy production and civilians" should be switched to "a potentially decisive offensive" against the U-boat fleet in the Atlantic.[55] Some in the Air Ministry labeled Blackett as a defeatist for his opinions, and that opprobrium extended to like-minded friends, including Hill. A new book, *Victory through Air Power*, written by Alexander de Seversky, a one-legged Russian flying ace from World War I, appeared in April. Seversky wrote some articles in the *Daily Telegraph* in which he criticized Hill, among others. Trenchard delighted in forwarding these to his friend, AV, citing it as evidence he was mistaken about the importance of bombing. Hill conceded that the campaign was having some effect, although this had been greatly exaggerated in the past. He finished on a conciliatory note, "I am very proud that you should think it worthwhile to convince me. . . . I go with you a very long way, but not the whole way. I know and honour the work you have done for the RAF."[56]

The next military catastrophe to occur was in North Africa in June where Rommel's Afrika Korps triumphed in North Africa over a dispirited Eighth Army. Churchill was in disbelief when he heard the news that Tobruk had fallen, an event he later described as a disgrace. He had forewarned General Auchinleck: "This is a business not only of armour but of will power."[57] The British tanks were no match for the Panzers, and Hill wrote to *The Times* that their deficiencies were due "to a system which has failed to anticipate future tactical requirements in guns, projectiles, armour, and performance; failed to collect, analyse, and profit by previous operation experience; failed sometimes even to obey the elementary rule that production must follow, not precede, development."[58] He repeated his arguments in more colorful language in the popular weekly *Picture Post*.[59] His main target was the policy of leaving weapon development to individual government departments that "tend to advertise their own wares, cover up their mistakes, shelter their boobies and avoid any degree of collaboration, external criticism or control which might limit their independence or sovereign rights."

Hill had become a fearless critic, not shrinking from strident public statements if he thought beneficial changes might result. He also began to display a more rapier wit in the House of Commons. In a short debate on women medical students, he asked the minister of labour, Ernest Bevin, whether he was aware that most London medical schools accepted no women students. Bevin of course was and suggested that as a member of the Committee on Medical Schools, Hill might take some action. Hill neatly sidestepped by asking, "Does the Minister consider it is a good thing to allow young men to stay out of the Services merely to prevent women entering the medical profession?"[60]

Hill continued his respectful disagreement about bombing strategy with Trenchard over the summer. In September, Trenchard arranged for Hill to spend a night in the lion's den—the home of Sir Arthur Harris, commander-in-chief of Bomber Command. Harris was notorious for his curt, dismissive attitude, so one can imagine how awkward the evening was for Hill. The two men stayed up until 2 A.M., and Harris confronted Hill with photographs of German cities devasted in the recent "1,000 bomber raids." Hill wrote to Trenchard about the visit, saying that while he sympathized with the many difficulties Harris faced, "I wish he would not exaggerate, and that he would state his case without bitterness towards other people. He would feel much better if only he had a clear idea of what our air policy is."[61]

Hill did not spend all his time mixing with senior figures in parliament or the Royal Society. Nor were all his concerns aired in public. As mentioned, his youngest son, Maurice, and nephew, Richard Keynes, both joined HMS Osprey, the anti-submarine experimental establishment, at Easter 1940. They were hired as temporary experimental officers and were not in navy service. Within months of taking up the position, Maurice wrote a blistering letter of complaint to his father about the chief civilian scientist at Osprey, B.S. Smith, saying that nobody "has any admiration or respect either for his executive or scientific prowess."[62] He also mentioned that their building had been straddled in a bombing attack on Portland dockyard so that there was hardly a window left and everything was covered in brick dust. Somewhat unkindly he added, "The only regret of the place is that the C[hief].S[cientist].'s office wasn't badly damaged and that he wasn't there when the bombs fell." The bombing raids caused the anti-submarine group to be relocated to Fairlie on the River Clyde in November 1940. Richard Keynes transferred there, but Maurice, frustrated by Smith's inertia, applied to join the Naval Research Establishment on the Firth of Forth. There he joined a group devising countermeasures to the German acoustic homing torpedo, eventually becoming the group leader.

Maurice wrote to his father again, early in July 1942, to tell him that his letter to The Times "evoked much sympathy" after being circulated at the Naval Research Establishment.[63] Although no longer working at the anti-submarine establishment,

Maurice had recently visited Fairlie for a meeting. His impression was the work there was proceeding extremely slowly, even though if it were to be successful, it could revolutionize naval tactics and "possibly stay the U-boat menace ... even in spite of the fact that U-boats are sinking times more tons ships than mines ever did, complete lethargy exists at the A/S establishment." B.S. Smith remained his *bête noire*. Maurice observed tartly that "Fowler, Blackett and Bullard have apparently failed in arousing any enthusiasm" for Smith's removal. Maurice was accompanied on his visit by the physicist Keith Roberts,[64] whom he regarded as the "greatest champion of the cause of the young and the temporary scientists." Roberts took charge of the Fairlie meeting after Smith warned that no answer to the U-boat menace could be expected for another year. Instead, it was agreed that new measures would be made available within twelve weeks.

Richard Keynes was even more acutely aware of the dispiriting effects of Smith's deadening leadership than Maurice. In August 1942, he and another temporary experimental officer courageously decided to speak out, in a way that the permanent staff could not. They offered their resignations in a letter of complaint to the director of scientific research at the Admiralty, which they placed on Smith's desk to be forwarded.[65] The letter said the majority of the scientific staff at Fairlie had lost confidence in the administration of the establishment and that made the effective conduct of research impossible. Smith was furious but, after days of bitter argument, did forward their letter to the Admiralty. Keynes knew that Maurice was on leave in Highgate so that he sent him a copy of the letter in the expectation that AV would become involved. A few days later, Richard received a letter from AV congratulating him and his friend on taking "the bull by the horns. You have done the right, patriotic and courageous thing."[66]

In fact, AV had already started to sound the alarm about Fairlie, based on what Maurice had told him weeks before. He drafted a note to the Third Sea Lord Frederic Wake-Walker, the controller of the Navy ultimately responsible for ships and equipment. Hill said that he had received a letter from a "young friend" not attached to Fairlie who had convinced him "it is certain that far less than the maximum result is being produced there."[67] He also told Wake-Walker that while he did not believe the War Cabinet knew about his concerns, he had discussed them verbally on 20 July with Sir John Anderson, Oliver Lyttelton (the minister of production), and R.A. Butler MP, who had taken over the chairmanship of SAC from Lord Hankey. Keynes was instructed by the Admiralty to obtain more evidence by interviewing scientists at Fairlie, and by early November Maurice was able to congratulate his "Dear Daddy," saying, "It's wonderful news from Fairlie, isn't it? I suppose you have been in large measure responsible for its sudden—even if belated—fulfilment."[68] Keith Roberts displaced Smith, who took his demotion badly, making the criticism at a staff meeting that the Admiralty was run by a clique of FRSs, trying to further their own interests, "the most foul deceitful, wicked men."[69] Maurice was amused by stories of scientists at Fairlie, who had detested Smith for years, crying crocodile tears over his downfall.

Meyerhof wrote a long Christmas letter to Hill in 1942, which, after family news, proceeded to an unofficial political briefing about three matters that concerned him: the Jews, France, and German intellectuals.[70] Although not a Zionist, he believed "the Jews still existing after the devastation in the zone of destruction" must be settled in a large part of Palestine "in spite of the Arabs." Referring to German academics, he observed that "they behaved most miserably and dishonestly and hypocritically and will continue to do so." In general, he thought Germany society was so poisoned that it could not be trusted to govern itself. Hill could not match Meyerhof's understandable ire, and replied, "I can't think Palestine would be capable of taking all the Jews who would like to leave Europe, and there are great difficulties in dispossessing Arabs and others and so exciting the animosity of the Mohammedan world."[71]

Meyerhof was able to continue his research, although the laboratory facilities at the University of Pennsylvania did not match those in prewar Heidelberg. He published an important paper in 1943 clarifying one of the fundamental reactions in the glycolysis pathway, where one molecule with six carbon atoms is cleaved into two monosaccharides or *trioses*, that each contain three carbon atoms.[72] He wrote Hill another long letter in September, reflecting on the dramatic recent progress in the fighting and the future of Germany:

> While I feel some mischievous joy in the smashing attacks of the RAF on Hitler's Berlin, my heart is bleeding for the destruction in Southern Italy ... where I have spent so much time and fruitful years working at the Zoological Station. And what will be destroyed, when the war climbs further north? Now it is indeed the time to discuss post-war problems.
>
> I would give Russia a greater share in the liquidation of Germany. ... Russia has even a greater interest than England to destroy the Junkers [landed gentry] and war potential.[73]

Meyerhof believed the Russians should occupy the eastern portion of Germany up to the River Elbe (which is where the US and Soviet armies did indeed meet up in 1945) so that they could "completely disband the Junker estates." He recognized that the removal of Germany's traditional ruling class might lead to "a labor and peasant Government" that should be accepted even if it meant communism, "provided that it is not a disguise for a new nationalism." He thought "that re-education must be mainly done by disaster, until the most stupid idiots understand that crimes and horrors do not pay—even University Professors; but re-education is hopeless, as long as the ruling classes are not completely annihilated." Meyerhof was optimistic that France would make a smooth recovery and even the Italians "may rediscover their soul in all ruins. ... The greatest problem remains Germany and the adjustment between Russia and the Allies."

As the balance of the war began to tilt favorably in 1943, Hill was able to relax from the sense of crisis that had driven him for nearly one decade, and his thoughts and activities slowly turned toward the future, as we shall see. Yet he remained committed to the primary objective of ending hostilities as quickly as possible. In a debate on the refugee problem in May 1943, he stated that no other action the Allies could take "to help the potential victims of Nazi misery can compare with what would be saved by shortening the war even by a fortnight." He continued:

> It is no use to make a quantitative estimate of the total suffering and loss in the world as a whole due to the present state of the war. The loss of life due directly to military action is only a fraction of that from other causes—famine, exposure, disease, disorder and massacre. That in its turn is only a small part of the total loss of human values—health, security, order, education, and prosperity—which the war has involved.[74]

Hill was now in his mid-fifties but retained his youthful physique—still running every morning on Hampstead Heath when he could. Margaret's energy levels matched his and she remained determined and optimistic throughout the war years. AV admired her ceaseless efforts in setting up Highgate Homes, offering shelter especially to the elderly who had been bombed out of their residencies. While Maurice was in Portland, Margaret put him in touch with old friends from Cambridge, the Pass family, who now lived on the Dorset-Devon border. Maurice announced his engagement to one of their daughters, Phillipa, in September 1943. AV's mother, crossing the street in Cambridge to mail a note of congratulations, was hit by a bus and died three days later. AV's sister, Muriel, a diabetic, had predeceased her two years earlier. Hill also lost two close friends to natural causes—William Hartree and Ralph Fowler[75]—but on a joyful note, daughter Janet gave birth to three war babies.

David Hill continued to follow in his father's traces. He spent some months as the resident scientist at A.A. Command at Stanmore, before moving to the War Office to serve under the Hill's close friend Sir Charles Darwin, now back from Washington and scientific adviser to the Army Council. At the end of 1942 David was sent to North Africa as a one-man OR unit embedded with the British First Army. He returned to England after about six months and in 1944 was selected by Brigadier Basil Schonland to be his personal assistant—Schonland knew David from Blackett's Circus days. In March 1944, General Montgomery chose Schonland to be his scientific adviser for the upcoming invasion of France. One of David's early tasks in France was to analyse the battle for Caen, which he reported to Schonland "had gone horribly wrong."[76] A bullet from a German sniper passed through the shoulder of his jacket while he was there. He spent the next year in France, Belgium and Germany, assessing the performances of the British and German land and air forces.

AV took special satisfaction from Gottfried Meyerhof's wartime work as a civil engineer. He was one of a small team responsible for the quality control of the materials and fittings for the huge, concrete Phoenix caissons that were essential components of the floating Mulberry Harbours used for the D-Day invasion.

Notes

1. O.F. Meyerhof to T. Meyerhof (15/10/40) MPT50 M613 UPenn.
2. O.F. Meyerhof to A. Makinsky (16/10/40), rockfound.rockarch.org/digital-library-listing/-/asset_publisher/yYxpQfeI4W8N/content/letter-from-otto-meyerhof-to-alexander-makinsky-1940-october-16.
3. R.A. Lambert memo "Dr Otto Meyerhof" (16/10/40) Rockefeller Archive Center, RG 1.1, series 200.A, box 113, folder 1390.
4. O.F. Meyerhof to G. Meyerhof (6/1/41) MPT50 M613 UPenn.
5. W. Meyerhof (2002), 56–67.
6. O.F. Meyerhof to G. Meyerhof (17/5/41) MPT50 M613 UPenn.
7. blog.nationalarchives.gov.uk/collar-lot-britains-policy-internment-second-world-war/ (accessed 22/10/20). Gottfried Meyerhof was one classed as "C" needing no restraint. He was working as a civil engineer and was not affected by the tighter restrictions.
8. E. Simpson to A.V. Hill (20/6/40) quoted in R.M. Cooper (ed.), *Refugee Scholars: Conversation with Tess Simpson* (Leeds, Moorland, 1992), 137–138.
9. A.V. Hill to A. Maxwell (24/6/40) quoted in Cooper (1992), 138.
10. api.parliament.uk/historic-hansard/commons/1940/may/23/enemy-aliens#S5CV0361P0_19400523_HOC_131.
11. E. Simpson to M. Schild (18/9/40) MS/835/14 Royal Society.
12. J. Black, "Heinz Otto Schild (1906–1984)," *Biographical Memoirs of Fellows of the Royal Society* 39 (1994): 383–415
13. Rall (2017) .
14. Beveridge (1959), 55–68.
15. Quoted in Rall (2017) .
16. Cooper (1992), 150.
17. M.H. Pirenne to A.V. Hill (21/3/42) MDA/A/24/4/36 R Soc Archives.
18. A.V. Hill, "Our Alien Friends," *The Spectator* (20/9/40) reproduced in Hill (1960), 231–232.
19. A.V. Hill, "Alien Internees," *Debate on the Address, House of Commons*, 3/12/40 reproduced in Hill (1960), 236–243.
20. *Daily Telegraph* 11/12/40 AVHL II 5/23. Hill seems to have overlooked the delays imposed by the censoring of letters from internees and the fact that mail from the Isle of Man passed through Liverpool, which was heavily bombed.
21. Letter (10/9/40) quoted in Angier I (1978):122.
22. Huxley (2003).
23. O.F. Meyerhof to G. Meyerhof (10/12/40) MPT50 M613 UPenn.
24. General Frederick Alfred Pile (1884–1976) commanded A.A.C. throughout World War II.
25. Minutes of Ballistics Committee, Ministry of Supply (9/8/40) AVHL I 2/3.
26. A.V. Hill to P.M.S. Blackett (12/8/40) PB/9/1/52/1940 R Soc Archives.
27. E.M.C. Clarke to A.V. Hill (29/10/40) AVHL I 2/3.
28. A.V. Hill, "Some Elementary Considerations (Not Commonly Realized) on AA Gunnery" (25/11/40) AVHL I 2/3.
29. E.M.C. Clarke to A.V. Hill (28/7/41) AVHL I 2/3.
30. Lord Beaverbrook to A.V. Hill (2/8/41) AVHL I 2/3.

31. A.V. Hill speech HoC (13/5/41), api.parliament.uk/historic-hansard/commons/1941/may/13/alien-doctors#S5CV0371P0_19410513_HOC_339 (accessed 10/22/20). This debate was at the end of the first day after the House of Commons chamber was destroyed by bombing on Saturday, 10 May. The Commons moved to Church House, Westminster. Three weeks before, incendiary bombs also caused major damage at UCL.
32. A.P. Brown, *The Neutron and the Bomb: A Biography of Sir James Chadwick* (Oxford: Oxford University Press, 1997), 195–213, and M. Gowing, *Britain and Atomic Energy* (London: Macmillan, 1964), 394–398.
33. Gowing (1964), 102.
34. Lord Cherwell to W.S. Churchill (27/8/41) quoted in Brown (1997), 218.
35. Gowing (1964), 106.
36. In a handwritten note to Hankey (2/8/41), Darwin asked the profound question: "Are our Prime Minister and the American President and the respective General Staffs willing to sanction the destruction of Berlin and the country round, when, if ever, they are told it could be accomplished at a single blow?" quoted in Brown (1997), 217. Darwin indicated that Roosevelt's senior scientific advisers, Bush and Conant, were prepared to discuss "whether the whole business should be continued at all." There is no evidence that Hankey shared Darwin's opinions with the SAC, and even if he had, they would probably have been regarded as beyond SAC's remit.
37. Gowing (1964), 106.
38. C. Andrew and V. Mitrokhin, *The Sword and the Shield* (New York: Basic Books, 1999), 150.
39. O.F. Meyerhof to G. Meyerhof (13/12/41) MPT50 M613 UPenn.
40. Gilbert (1991), 712–715.
41. A.V. Hill, "Combined Operations and a Great General Staff," undated and unsigned memo AVHL I 2/1.
42. M.J. Nye, *Blackett: Physics, War, and Politics in the Twentieth Century* (Cambridge, MA: Harvard University Press, 2004), 76–77.
43. P. Joubert de la Ferté to A.V. Hill (20/2/42) AVHL I 2/2.
44. api.parliament.uk/historic-hansard/commons/1942/feb/24/ministerial-changes#S5CV0378P0_19420224_HOC_256.
45. A.V. Hill and R. Glyn Letter to *The Times* (23/2/42) AVHL I 2/2.
46. P. Joubert de la Ferté to A.V. Hill (4/3/42) AVHL I 2/2.
47. A.V. Hill to P.J. Grigg (27/3/42) AVHL I 2/2.
48. P.J. Grigg to A.V. Hill (3/4/42) AVHL I 2/2.
49. A.V. Hill to P.J. Grigg (10/4/42) AVHL I 2/2.
50. R.M. Barrington-Ward to A.V. Hill (17/4/42) AVHL I 2/2.
51. P. Joubert de la Ferté to A.V. Hill (1/5/42) AVHL I 2/2.
52. api.parliament.uk/historic-hansard/lords/1942/may/05/organization-for-joint-planning.
53. Viscount Trenchard to A.V. Hill (23/2/42) AVHL I 3/96.
54. T. Wilson, *Churchill and the Prof* (London: Cassell, 1995), 74.
55. Nye (2004), 80–82.
56. A.V. Hill to Viscount Trenchard (8/6/42) AVHL I 2/2.
57. Gilbert (1991), 721.
58. A.V. Hill (1/7/42) Letter to *The Times*.
59. A.V. Hill, "A Practical Plan to End Our Military Weakness," *Picture Post* (18/7/42) AVHL II 1/12. The sub-editor employed the old Fleet St trick of using gigantic font to emphasize the frightening phrase, "OUR MILITARY WEAKNESS."

60. api.parliament.uk/historic-hansard/commons/1942/aug/06/medical-schools-london-women-students#S5CV0382P0_19420806_HOC_8.
61. A.V. Hill, "Trenchard, an Argument" M&R: 198–203 AVHL I 5/4.
62. M.N. Hill to A.V. Hill (15/9/40) AVHL II 5/31.
63. M.N. Hill to A.V. Hill (6/7/42) AVHL II 5/31.
64. John Keith Roberts FRS (1897–1944), Australian physicist. Trained under Rutherford in Cambridge; responsible for developing techniques to degauss ships while in charge of the Naval Research Establishment.
65. I. Glynn, "Richard Darwin Keynes (1919–2010)," *Biographical Memoirs of Fellows of the Royal Society* 57 (2011): 205–227.
66. A.V. Hill to R.D. Keynes (16/6/42) KEYN 1/001 CAC.
67. A.V. Hill to Vice-Admiral F. Wake Walker undated, handwritten note AVHL II 5/31.
68. M.N. Hill to A.V. Hill (2/11/42) AVHL II 5/31.
69. M.N. Hill to A.V. Hill (9/11/42) AVHL II 5/31.
70. O.F. Meyerhof to A.V. Hill (5/12/42) MDA/A/26/11/69 R Soc Archives.
71. A.V. Hill to O.F. Meyerhof (1/2/43) MDA/A/26/11/68 R Soc Archives.
72. Needham (1971), 121.
73. O.F. Meyerhof to A.V. Hill (25/9/43) AVHL I 3/58.
74. A.V. Hill "The Refugee Problem" House of Commons (19/5/43) reproduced in Hill (1960), 256–62.
75. Fowler continued his prodigious wartime service both on the Ordnance Board and working with Blackett at the Admiralty after returning from Canada in 1941. He was knighted in 1942. After his death in 1944, Hill wrote: "He gave all he had. ... For four and a half years he drove and coaxed the failing machinery along, perfectly aware that it was failing, bravely and cheerfully till it stopped." *The Times* (5/8/44).
76. Huxley (2003).

12
Restoration

In March 1941, President Roosevelt dispatched a trio of scientists to England, led by James B. Conant, the president of Harvard and a member of the National Defense Research Committee. They stayed for six weeks, gathering information about the utilization of scientists and engineers in the war effort and establishing an NDRC office at their London embassy to encourage technical collaboration. Conant was fêted—lunching with Churchill and Lindemann, meeting the king, and honored with a large reception at the Royal Society (RSoc). Hill was instrumental in arranging Conant's election to foreign membership in May, sending a cable: "Greetings foreign member Royal Society, but not very foreign."[1] In an article for *The Times*, Hill referred to Conant's visit and the burgeoning Anglo-American co-operation that had developed since the Tizard mission. In his concluding paragraph, he looked forward to the continuation of exchanges after the war: "The problems of reconstruction are bound to be largely scientific, and it will be necessary to start up again one day all the peaceful scientific enterprises."[2] It is noteworthy that he took such an assured, long-term view publicly—six months before Pearl Harbor brought the United States into the conflict, and five days before the Germans terminated their non-aggression pact with the Soviets by launching Operation Barbarossa.

In the same confident spirit, a number of science events were held in war-ravaged London. The most diverse, perhaps, was a conference held in the historic lecture theater of the Royal Institution on "Science and World Order" in September 1941. It was initially planned to take the place of the BAAS annual meeting, but soon turned into an international affair with scientists from more than twenty countries attending (many courtesy of SPSL). There was a celebratory lunch at which Foreign Secretary Anthony Eden spoke. Each of the six sessions of talks over three days was chaired by a distinguished figure, including the American, Chinese, and Soviet ambassadors and H.G. Wells. His own talk had been cut off for running over so that when Wells took control of the final session on "Science and the World Mind" he was in "a petulant temper . . . and determined that no one else should have a second of extra time."[3] The statistician and fellow science popularizer Lancelot Hogben found himself being pulled down by the seat of his pants as soon as the warning red light flashed! Hill, who was on the organizing committee, spoke in the first session on "Science and Government" and argued for preserving the independence of science, a stance considered outmoded by radical colleagues

Bound by Muscle. Andrew Brown, Oxford University Press. © Oxford University Press 2023.
DOI: 10.1093/oso/9780197582633.003.0012

such as J.D. Bernal. Hill warned that the authority, routine, secrecy, discourage-
ment of initiative, and lack of freedom and of criticism inherent in bureaucracies
"provides the antithesis of the environment in which good scientific work is usually done."[4] Over the three days, Hill heard too much of the phrase "dialectical
materialism" and too many calls for the abolition of capitalism after the war for his
liking, but he was encouraged by the unanimous opinion, frequently repeated, that
"victory over aggression throughout the world is a necessary preliminary to any
reconstruction."

The conference garnered much media interest, and Hill took part in a small group
discussion with a quartet in New York that NBC arranged for the benefit of American
radio listeners. He made several other wartime broadcasts at the invitation of the
BBC. In April, he addressed scientists in Germany and started by telling them that his
friend Meyerhof was now safely in Philadelphia.[5] He was also invited to reply to Peter
Kapitza, who had sent a radio message from Moscow after the Germans attacked the
USSR. He did not know how many Russians were able to hear his talk but was de-
lighted to learn later that it was picked up in southwest France by physiologists in the
Resistance.[6]

We have seen that Hill was very active in the House of Commons throughout
1942. He was present on 17 December 1942 when Eden read a text from the United
Nations[7] stating that the Nazis were "now carrying into effect Hitler's oft repeated
intention to exterminate the Jewish people in Europe."[8] After short, dignified
speeches by several Jewish MPs, the "the House as a whole rose and stood for a few
frozen seconds."[9] In a follow-up session one month later, Hill asked the home secretary, Herbert Morrison, whether any policy change had been made on the issue
of visas to refugees from enemy or enemy-occupied countries. Hill was joined by
the redoubtable Eleanor Rathbone MP, who then harried Morrison to the extent
that he rebuked her for suggesting the British authorities had shown "extreme mea-
gerness" in the number of visas issued.[10] At a business session in March 1943, Hill
requested that time be made to debate the motion standing in his name and the
names of 276 other MPs that maximal efforts should be made to offer help and temporary asylum to "persons in danger of massacre" who managed to escape from
Nazi-occupied Europe.[11] A debate on the refugee problem took place in May, again
with Rathbone playing a prominent role and not shying away from disputes with
ministers. She understood that there were many competing demands on the minds
of senior government figures, but she was responding to the constant stream of letters "from agonised people who feel that the one chance left for their relatives is
slipping from them and that they may soon have to take that awful journey to the
Polish slaughter-house and who beg me to rescue them, not realising how impotent
I am."[12] In his speech, Hill praised the Hon. Lady as "the patron saint of refugees."
Hill said he could not match the moving quality of Rathbone's speech or that of
Colonel Cazalet, who had just informed the House of recent intelligence received
from Poland about a camp for the extermination of Jews named Treblinka. Hill

promised to be colder and more arithmetical by comparison. The nub of his argument was delivered succinctly:

> The loss of life due directly to military action is only a fraction of that from other causes—famine, exposure, disease, disorder and massacre. That in its turn is only a small part of the total loss of human values—health, security, order, education and prosperity which the war has involved. In Europe as a whole the civilian death-rate may very well be increased by a half in the war. That is probably a moderate estimate. That would mean 3,000,000 or so extra deaths per annum. Adding those in China and in countries now occupied by the Japanese, and including direct military losses, I imagine that in the world as a whole there are between 5,000,000 and 10,000,000 people dying annually owing to the war, that is, between 100,000 and 200,000 per week. This is altogether apart from the other losses in human values.
>
> The only way to save those lives and those values is to bring the war as soon as possible to a victorious end. Nothing we can conceivably do otherwise to help the potential victims of Nazi misery can compare with what would be saved by shortening the war even by a fortnight.[13]

Remarkably for a man who had spent hundreds of hours assisting refugees from Europe through SPSL, Hill thought any impact that further efforts could achieve would be "pitiably small." Yet, for precisely that reason, help should be offered quickly and efficiently since it would not interfere with any other war efforts. Lastly, he addressed the issue of anti-Semitism, which other MPs had warned would result from an influx of Jewish refugees. Describing it as the argument of the last straw, Hill observed the arrival of another 10,000 Jewish refugees would equate to one in 5,000 of the British population. To suggest, as responsible people sometimes did, that this would produce a serious risk of anti-Semitism "is a gross insult to the intelligence, good nature and commonsense of the normal citizen and is to confess oneself the foolish dupe of Nazi propaganda."

Meyerhof wrote to him at the end of 1942 complaining about naïve overoptimism in the United States on postwar reconstruction, and Hill agreed, "The war isn't over by any means yet."[14] But as the momentum began to shift in the Allies' favor, Hill became ever more focused on his return to academic life and especially on the restoration of science in Great Britain and beyond, after the end of hostilities. By 1943, Hill was planning to resume his career at UCL as a Foulerton research professor funded by the Royal Society. Many of his friends and colleagues wanted him to stay on as an MP as well. Hill was dubious about the possibility of such a double role, although J.M. Keynes was one voice strongly in favor of it. Hill discussed the issue with the master of Magdelene College, A.B. Ramsay, who was also chairman of the University Conservative Association. Ramsay said he would not press him to remain as an MP although everyone regarded him "as an ideal University Representative."[15] Hill confided in his fellow RSoc secretary, Jack Egerton, that he had no wish to and, in any case, could not afford to give up his professorship at UCL if it were renewed, so it would be improper to carry on as an MP.[16] The matter was finally settled by the president of RSoc, Sir Henry Dale, who agreed with Hill that he could not continue as a Foulerton professor and as an MP after a general election. Pointing out that monies were given to the RSoc "for the direct support of research, and not to maintain a science advocate in Parliament," Dale did not want to change the rules or to compromise. He added, "Your success as a champion of science among the politicians is no longer a matter of surmise, and your retirement will entail a heavy loss to an important cause, at a critical time."[17] Hill was grateful to Dale for his unequivocal opinion, saying it was exactly what he needed "to make the matter clear to my Cambridge friends."[18] Hill forwarded Dale's letter to Ramsay, who agreed that it made it "quite clear that you cannot do what Keynes thought would be possible."[19] He asked Hill, as before, to stand if there was a snap general election, otherwise to give early notice of resigning the seat.

In February 1942, *Hansard* records[20] that Professor A.V. Hill asked the secretary of state for India, Leo Amery, "whether he is aware of the close and profitable collaboration in scientific research and technical development now existing between the United Kingdom, the United States of America and the Dominions, in the common war effort; and whether steps have been taken to establish similar relations with the distinguished scientists, and the great scientific and technical resources of India?" Amery laid out the bare details of existing collaboration, before agreeing with Hill that more personal contacts would be helpful. Hill and Egerton had already set up a British Commonwealth Science Committee at the RSoc to promote scientific research amongst the various countries, especially as applied to technical, biological, medical, and economic problems.

The committee issued its first report in March 1943, by which time the situation in India had deteriorated all around. The Japanese had occupied Burma, Churchill was resisting increasing pressure to promise India independence after the war, Gandhi and other political leaders were interned, and, most tragically of all, a famine was unfolding in Bengal that would eventually kill three million. In the early summer of 1943, Amery informed Hill that the Indian government wanted him to spend time in India, assessing the state of scientific research and advising how it might be harnessed for development in the future. Hill's first reaction was to consult a fellow MP, Sir Stanley Reed, who had edited *The Times of India* for decades, and "his advice was emphatic."[21] Reed told Hill that if he went as a representative of the British government, he would be received with general mistrust; instead he should arrange to travel under the auspices of the RSoc.

Amery agreed with the indirect approach and arranged for General Wavell, the Viceroy, to make the formal request to the RSoc.[22] As the proposed visit took shape over the summer, Hill suggested that he should extend the visit by continuing on to China after India. Amery could see no objection after discussing it with Sir John Anderson, but suggested that Hill should wait to see how the Indian government felt about it.[23] Prime Minister Churchill furnished Hill with a letter sending his good wishes to "Indian men of science" noting that through a "monstrous perversion" the fruits of science had been turned to evil ends, but agreeing that when the war was won scientists would play a role in "directing knowledge lastingly towards the purposes of peace and human good."[24] Hill left from Poole on a flying boat, making stops in Ireland, Lisbon, Gibraltar, and Djerba Island before reaching Cairo on the 11 November. He was able to spend three days of relaxation at the Heliopolis Hotel, managing to see the Pyramids and the Sphinx. His plane then hopped to the Dead Sea, Bahrain, and Dubai before crossing the Arabian Sea to Karachi, Gwalior, and finally Delhi.

Hill's tour of universities, hospitals, schools, and factories was arduous: he swore that he shook hands with several thousand people. Hill's natural openness and friendly nature allowed him to quickly gain the confidence of new acquaintances. His energy and work ethic remained undiminished—he gave lectures, compiled voluminous notes, and broadcast on all-India Radio. He visited one dozen research centers, traveling thousands of miles. He appreciated how isolated the scientists in India felt, and although he might have felt embarrassed at the number of talks he was invited to give, he felt he "must do what was physically possible."[25] He also attended various conferences, the highlight being the Indian Science Congress. Hill had the idea to bring a sheet of vellum with him so that four distinguished Indian scientists[26] who had been elected to foreign membership but had not been able to come to London could sign it. It was the first meeting in the 300-year history of the Rsoc to be held outside England. By emphasizing the RSoc rather than his position as an MP, Hill seems to have avoided any political friction and was given every possible encouragement and assistance. Sir Shanti Bhatnagar (Figure 12.1), who obtained his doctorate at UCL and was now the founding director of the new Council of Science and Industrial Research, urged him to criticize openly whatever he thought was "wrong or stupid."[27]

Another old friend, Dr. Sohan Lal Bhatia, provided him with hospitality whenever he was in New Delhi. Bhatia was a medical student at Cambridge before World War I, and Hill got to know him in 1911 when he was demonstrating in physiology classes. After qualifying in London, Bhatia served in the Indian Army and won the Military Cross for tending the wounded under heavy artillery fire. After World War I, he returned to India and became a professor of physiology in Bombay; by the time of Hill's visit he was deputy director of the Indian Medical Service and a major general. He informed Hill about the "sorry" state of medicine and medical research in India, and together they hatched the idea of establishing an All-India Medical Centre.[28] Hill had just completed two years on the Goodenough Committee reviewing the state of British medical schools, so he was familiar with the challenges of improving medical education. Its seminal report suggested fundamental changes to medical education in the United Kingdom to provide for the anticipated National Health Service.[29] Hill viewed "better health as the first need of India" and cited malaria as public enemy No. 1. As he wrote in his report to the Indian government,[30] the intention of the Centre would be to produce the future leaders of Indian medicine and public health, the teachers and the research workers. Selection should be based on merit, regardless of race, religion, politics, or background; staff should be paid an adequate salary and

Figure 12.1 Hill and Sir Shanti Bhatnagar FRS, "the father of India's research labs," 1942
Photograph courtesy of Professor Nick Humphrey/Churchill Archive Centre, Cambridge (Papers of A.V. Hill, AVHL).

students supported by generous grants. He emphasized that it should be adminis-tered as an all-India affair, free from provincial and inter-state rivalries, and partly for those reasons it should be situated in the capital Delhi. He was also in favor of creating a central organization for scientific research—something he would surely have objected to in the United Kingdom. He recommended that the whole hierarchy would be overseen by a minister for planning and development with input from a consultative committee of scientists and industry leaders. Below that there would be six national boards, equivalent to the MRC, DSIR, and others in Britain, that would oversee research in medicine, industry, agriculture, natural resources, engineering, and war. By the end of 1944, the government of India had adopted many of his suggestions. A Scientific Consultative Committee was appointed and laid the groundwork for five national laboratories that appeared over the next few years.[31]

Hill's other major push was to end the international isolation of Indian research and, with the approval of Viceroy Wavell and Secretary of State Amery, he planned to bring a party of scientists back to England with him. He also recommended that young Indians came to England to study, and as soon as possible because there would be a huge demand from British candidates when the war ended.

The India Office suggested that Hill should offer his services to the South East Asia Command.[32] The Supreme Commander of SEAC, Lord Louis Mountbatten, had been used to working closely with Desmond Bernal and Solly Zuckerman in England, before his recent appointment. During Hill's five-month stay, heavy fighting broke out in Burma involving the British and Indian armies. Hill began to address the health risks and other physiological challenges facing troops fighting in the tropical jungle. As a staunch proponent of operational research, he naturally recommended that a scientific adviser be assigned to the commander-in-chief. Mountbatten was delighted. Hill contacted Sir Henry Dale to say he was going to cut out his proposed visits to China and Kabul because Mountbatten had asked him to stay on in India to advise on scientific research within SEAC and to provide scientific liaison with England.

Jack Egerton wrote to Hill while he was in India to inform him of one especially significant development at the RSoc. About a year earlier, J.B.S. Haldane FRS, the geneticist son of Hill's old Oxford friend, had written an article advocating that women should be eligible for election. Hill wrote to him at the time saying that the RSoc did not exclude women, and Haldane was perfectly at liberty to nominate a woman for fellowship. He added: "Personally I should be very glad to see a woman elected, provided, of course, that she was elected on merit." In his opinion, Honor Fell, a pioneer in tissue and organ culture at the Strangeways Laboratory in Cambridge, would "have a fair chance" of election.[33] By the end of 1943, as Egerton explained:

> There has been a lot of a flutter, as two certificates have come in for women candidates: we realize (as you said in reply to JBS Haldane) they *can* be selected, but I don't think I quite realized to what extent the law was mandatory as a result of the Sex Disqualification Act.

The RSoc had taken legal advice, which was that its statute needed to be updated to bring it into line with the law that had been passed in 1919![34]

Hill's desire to return to laboratory life was reinforced by his time in India. He wrote to Margaret from New Delhi, "I don't want to be a public character and want to crawl out of being a 'figurehead' and an MP and a jolly old busybody as soon as I can."[35] The political status of India had been a contentious political issue for years. The India Act in 1935 introduced some element of self-government, and by 1944 there was general acceptance in the British parliament that there would be some form of independence after the war. Hill gave his most substantive political speech when the future of India was debated in July. Although there was an air of compromise in the House, major questions such as whether India would remain a single country or be partitioned into two or more nations were undecided. Hill prefaced his remarks with a warning that he believed "India was living on the edge of a precipice."[36]

> The factor of safety is so low that any disturbance, even a comparatively minor one, may send her over the edge. For that reason, we must regard this not as a matter which can be thought out slowly; it is not one in which time is on India's side. It is a matter of great and extreme urgency. . . . I have grave doubts about the wisdom of us urging from here that any consideration should be given to partition. Devolution, yes; self-government, like we have in this country, within limited regions, yes; but partition, in the sense of having five separate Dominions, or whatever it may be, in India could, I think, only lead to what I myself in my less polite moments refer to as the Balkanisation of that great peninsula.

Hill's concern about the potential instability of India did not stem as much from the existing communal or religious differences, which he thought had been exaggerated by some Indian politicians, but because "the factor of safety in India is almost zero."

> Disease and malnutrition, working together, produce a vicious circle, making a situation so near the margin that any internal strife and disorder on the one hand, or any serious epidemic, like that of 1918, on the other, might produce a major catastrophe. Yet in spite of this, the population of India is increasing by 15 per 1,000 per annum, or about 6,000,000 a year. This, it is necessary to emphasize, is no new thing. The Indian population has always been living right up to its income, in the matter of health and food. If health measures are improved, and food production and distribution bettered, then this 6,000,000 will, as has been said, become 7,000,000, 8,000,000 or 9,000,000 per annum. How can food supplies catch up and keep pace with so riotous an urge to reproduce, particularly in a population which is living for the most part in poverty, and not infrequently in misery, and is so ill-educated that even to-day only about 8 per cent of the female population of India over 5 years of age can read and write? Many of these things will depend mainly for their solution on the women. That is the real problem of India as I and many others see it. It depends upon the six terms—health, food, population, agriculture, poverty and education. That is the sort of problem which will not yield simply to political dialectic, or to the manufacture of political machinery. It requires complete and deliberate co-operation all round, hard thinking and hard work.

Encouraged by a number of recent developments, Hill believed "India is ripe for a great technological development of all her resources. I can see little hope for India of greater prosperity apart from going with the stream of modern life and seeking her prosperity in that kind of development. The essential condition for success is a reasonable degree of economic and political unity."

On his return to London, Hill found a concerted effort within the RSoc under the rubric of "Post War Needs." He attended his first meeting of the biology committee in early May; the minutes from that meeting, aside from the expected calls for more generous government funding, emphasized the need for nature reserves as places to study flora and fauna under field conditions.[37] Similar committees considered the future needs of chemistry, geography, geology, geophysics, and physics. Egerton and Hill analyzed each committee report, with Hill offering his "frank opinions" on various drafts to President Dale.[38] The final, general report was circulated among fellows in January 1945. Regarding biology, the need to develop biochemistry as a research area in UK universities was stressed. Hill, in addition to his general contributions, was determined to promote the future of biophysics. He was successful in securing a substantial grant from the Rockefeller Foundation to restart work at UCL. In December he circulated a memorandum, "The Need for an Institute of Biophysics,"[39] to Dale and to Sir Edward Mellanby, the secretary of the MRC, among others. Although he believed such an institute should be sited in London, the cost of it would exceed the capacity of usual university financing and should be met by the RSoc making a direct approach to the Treasury. He elaborated on modern technology now available for research including radioactive isotopes, electron microscopy, and other laboratory electronics that could be applicable in biophysics. He foresaw that developments in genetics offered "a wonderful field for fine physical technique." He warned that the plan would fail if not attempted on a sufficiently large scale. At the age of sixty, worn out by his wartime efforts, he would not be interested in directing the proposed institute himself but wanted to return to his own laboratory. In addition to pleading the case for biophysics, Hill called attention to rare subjects and possible areas for research that fell between or overlapped the major scientific disciplines.[40] He persuaded the existing RSoc sectional committees to come up with ideas—which they did on a multitude of topics, ranging from algae and analytical chemistry via computers and statistics to epidemiology and virology.

In the summer of 1944, the Foreign Office began to ask some émigré scientists about the Nazi sympathies of their former colleagues in German universities. The first to consult Hill about it was Herbert Herxheimer, a Jewish chest physician with an interest in sports medicine, whom Hill first met at the Amsterdam Olympics in 1928. He had come to London in 1938, and Hill secured him an appointment as the physician to Highgate School. Herxheimer was worried that the Foreign Office inquiry could lead to "incomplete and haphazard results."[41] Hill agreed "unless the work were done with the utmost discretion and care, the result might be very unsatisfactory."[42] Hermann Blaschko, who had recently transferred from Cambridge to the pharmacology department in Oxford, also wrote to Hill for guidance. He again urged caution: "It is not a simple matter, a great many German scientists ... may have found it necessary to conform ... people can be deceived and misled by propaganda ... compile your replies with utmost care."[43] In July, Meyerhof wrote him a seven-page letter,[44] with the following passage underlined in red pencil:

> I think one must support Russia and all the weakened neighbour states to keep Germany down, economically and politically and try any kind of re-education only when no power is left . . . the Universities should be closed for a long time.

Meyerhof continued in uncompromising fashion:

> One cannot overlook that these gas-chambers, in which the Nazis kill daily 5,000 jews since over 2 years, are constructed by chemists, engineers and physicists, educated in university and there is a continuous line from the jew baiting and pogroms, inspired by students and academic people until this fiendish and devilish mass extermination. I am not blind against the other achievements of the German Universities, but without a radical dissolution of what is left and a complete re-education in a much later time, the poison will not be eliminated.

Meyerhof wrote his letter from the Woods Hole Marine Biological Laboratory on Cape Cod, where he and Hedwig liked to spend their summers. He included family news and briefly mentioned that he expected to start research on penicillin at the request of A.N. Richards in Philadelphia. He did not tell Hill that he had sustained a heart attack at the end of June after a tennis game and was on strict bed rest, as ordered by a medical friend from New York. His son, Walter, came to see him and found him so weak that he could hardly speak. Otto was being nursed by Hedwig but experienced further cardiac problems. At the end of the summer, he was transferred to Mount Sinai Hospital in New York, where he remained for ten months.[45] Meyerhof and his wife were finally able to return to Philadelphia just after the war was over; Walter was struck by how much he had aged. Otto wrote to AV in September 1945, saying he had just returned to his laboratory, but his brain was rusty and his body weak.[46]

The main purpose of this letter was to ask Hill "how one should proceed with our former German colleagues." The US Army occupied Heidelberg in April and closed the university in order to consider how to cleanse its Nazi associations. A SHAEF intelligence report in May judged the professoriate to be generally "pro-Nazi" while the students were "mildly and moderately Nazi."[47] This assessment was in stark contrast to a powerful myth developing that academics never wanted anything to do with Nazi ideology (and even if they had joined the NSDAP, it was under coercion); they were essentially victims of the violent fanatics who had corrupted the tolerant spirit that reigned before 1933. Although the intelligence officers with the greatest insight doubted whether the university could be trusted to oversee its own denazification, it was permitted to reopen in January 1946. Meyerhof told Hill he had received "a somewhat evasive letter" from Richard Kuhn, the director of the KWG institute in Heidelberg who had undermined his status in 1936. Kuhn now took credit for saving Otto Warburg from "difficulties with the Nazis." He claimed to have done this by interceding with the office of the "Führer." Meyerhof noted his disgust in the margin of his letter to Hill: "*He call this sadistic crank and dead rat still 'Führer'.*"[48] Hill's response was uncharacteristically bland:

> I am afraid that Kuhn is somewhere between betwixt and between. No doubt his fundamental sympathies were anti-Nazi but one gathered that at a Chemical Congress just before the war, he was the leader of an enormous party of Germans and gave the Hitler salute and said "Heil Hitler" as he opened the proceedings at the Congress. I suppose he felt he was taking out an insurance.[49]

Kuhn retained his directorship of the Heidelberg institute and wrote again in October, offering Meyerhof his old position back as head of the physiology department.[50] Replying to Kuhn, Meyerhof explained it was no longer easy "to write to you openly in the spirit of our old comradeship." He did not mince words and accused Kuhn directly "of having freely placed his remarkable scientific ability and mastery of chemistry at the service of the regime whose unspeakable heinousness and infamy" was "well known" to him.[51] The betrayal was particularly painful to Meyerhof because he was aware of Kuhn's "liberal spirit." Regardless of any future decision, Meyerhof would now insist on the payment of severance and full restitution for the loss of his property (especially his library). Kuhn previously informed Meyerhof that his personal property had been confiscated at the end of 1939. His loyal lab assistant, Walter Schultz, was trying to track it down for him and even bought back Meyerhof's pictures at auction. Schultz would not accept any reimbursement because Meyerhof had been so kind to him, but it was suggested that Meyerhof might send him a food parcel.[52]

Toward the end of 1946, Meyerhof was asked to give his opinion on Kuhn by the American authorities overseeing Heidelberg University. He believed that Kuhn "was not politically minded" but had "sided with the Nazi regime in some important matters." Meyerhof felt as though he had been a restraining influence until he left in 1938, by which time Kuhn believed the regime "was irrevocably entrenched in power [and] he was ready to compromise his great scientific reputation without scruples. I am convinced that he did this from expediency and from weakness of character and he never held any Nazi convictions." Unlike many ex-colleagues in Heidelberg, Meyerhof rejected any excuses or false rationales that Kuhn offered and recommended that he be allowed to continue a research career but removed from all leadership positions in German chemistry and not be "entrusted with the education of University students."[53]

Hill did not stand again as a candidate in the general election of July 1945, which was won in a landslide by Attlee's Labour Party. Shortly after the result, he wrote to

Meyerhof: "Churchill has been a magnificent war leader, but he is not suited to peacetime conditions."[54] Ironically, the internal politics of the RSoc gave Hill no respite. Three out of the five senior officers, the president (Dale), foreign secretary (Tizard), and Hill, were due to stand down at the end of 1945, to be replaced in a seamless, gentlemanly transition. The officers were all members of the controlling Council of the Society, which in total consisted of twenty-one Fellows. Every year there would be a turnover of about half the Council, but the senior figures remained for the terms of their office (usually five years). All these positions were to be filled after elections involving the Fellows on St Andrew's Day, as prescribed by the Royal Charter of King Charles II. A physics professor from UCL, Edward Neville da Costa Andrade, known to his chums as Percy, finished a two-year term on the Council in November 1944. He was actively involved in the physics committee on Post War Needs and felt the next president needed to be politically astute in order to make the RSoc a vital cog in the machinery of national science policymaking. After informal conversations with Fellows and Officers (including Hill), he decided to submit his ideas to the Council in the form of a "memorial."[55] Andrade claimed that Hill was acting "as a wise friend and not as an official of the Royal Society," whereas in his decade as biological secretary Hill had always been fastidious about considering his official position above any personal opinions.

Andrade started to circulate his draft memorial early in 1945 and was heartened by Hill's reaction. He wrote to an ex-president Sir Charles Sherrington in February that Hill "considers our action a perfectly proper and constitutional one, welcomes it and thinks we are doing the Society a service. . . . If A.V. Hill thinks that all is well I do not think there can be much wrong."[56] Andrade had also communicated with the man he favored as the next president, Sir Henry Tizard. Tizard, of course, was well-versed in the ways of Whitehall, but he did not meet the unspoken tradition of being a preeminent scientist, one who had been awarded the Copley Medal, for example. There were two early candidates, Robert Robinson and G.I. Taylor, who both met the standard as scientists but neither of whom was greatly interested in science policy. Andrade submitted his memorial, signed by eighty-four Fellows, to the Council in March; Tizard was appalled to see his own name cited as an ideal candidate for the presidency. He demanded that it should be removed, or he would resign from the Council. Hill and Egerton sent a letter to those who had signed the Andrade memorial explaining that Tizard would exclude himself from any further consideration unless this was done. Andrade's memorial did have the immediate effect of expanding the possible field of presidents. Hill's name was one of several added to the list, strongly favored by Tizard, who now agreed to his own name going forward. Hill, explaining that he just wanted to return to the laboratory, withdrew his name in May, followed by the apparent favorite, Tizard, in June. Hill was disconsolate about Tizard's exit from the race and told him, "The more I see of politics, the more I like science."[57] Hill did agree to succeed Tizard as foreign secretary of the Society. Dale, saddened by developments, wrote to Hill in quasi-Churchillian tones: "Never, in the long history of our Presidential elections, has so much mischief been caused for so many by so few."[58] Hill was furious with Andrade for his meddlesome campaign.[59]

> Dear Andrade,
>
> I think you ought to know, if you haven't heard already, that as a direct consequence of the memorial, and of the reaction it produced in some quarters, Tizard has absolutely refused to have his name considered.
>
> Had it not been for the memorial it is very likely that he would have been nominated.
>
> Partly as a result of it, two other people have refused to have their names considered.
>
> To Andrade E N da C: mix two metaphors—the milk has spilt & the least said now the soonest mended.
>
> Yours sincerely,
> A.V. Hill

After all the intrigue, Robinson beat Taylor by one vote in Council and became PRS on 30 November. As Andrade and others feared, the cantankerous Robinson avoided science policy and public affairs and was always happiest talking about organic chemistry.[60] It was customary for officers of the RSoc to be honored with a knighthood. When it was Hill's turn in 1941, he declined the award—much to Margaret's annoyance. It seems that he was motivated by more than not wanting to be called "Sir Archibald." "It will be a sad day for British Science," he noted, "if the Officers of the Royal Society come to be regarded as officials of the Department of Education and Science or of Trade and Industry and are expected to take orders from Ministers." He satisfied himself by listing the names of some greats who lived and died without honorific titles: Darwin, I.K. Brunel, Huxley, Faraday, and Maxwell.[61] Sir Henry Dale, shortly before leaving presidential office, wrote to Prime Minister Attlee expressing concern that Hill was about to leave public service without any recognition. He asked Attlee whether he might recommend him for the Companion of Honour.[62] The high award carries no title and so was acceptable to Hill. After the war, he also accepted the US Medal of Freedom and the French Légion d'Honneur.

In May 1946, Hill spent several weeks in North America, flying both ways across the Atlantic. He stayed with the Meyerhofs in Philadelphia—it was an emotional reunion for both men who had not seen each other since 1937. Hill was pleased to attend their swearing-in ceremony as American citizens. On his return to London, he plunged into the RSoc Empire Scientific Conference, which lasted for three weeks and was immediately followed by the delayed celebration for the tercentenary of Newton's birth. As he discussed with Meyerhof, a special invitation for the Newton meeting was sent to Max Planck, the eighty-eight-year-old German physicist who embodied the ethos of an earlier period. His son had been tortured and hanged for his role in the July plot on Hitler's life. Meyerhof expressed concern about Planck's physical and mental frailty, but Hill reassured him that Col. Blount, who would be bringing him from Germany, said "the old chap is really all right."[63] Col. Bertie Blount was an Oxford-trained chemist who had worked in Germany in the 1930s and enjoyed an adventurous war in the Intelligence Corps, inventing equipment for the Special Operations Executive. He was now the officer in charge of research and science for the British occupation zone; in 1945 he had been responsible for extracting Planck from the Soviet zone to live in Göttingen. Blount told Hill that inviting Planck was seen as "a gesture of fairness and goodwill" in Germany, easing his relations with the science community.

The chemist Otto Hahn had been appointed as the new president of KWG in April, and at the end of the year he was scheduled to go to Stockholm to receive the Nobel Prize for his role in the discovery of nuclear fission. Meyerhof wrote to congratulate him on both achievements:

> The boundless disaster in Germany, which to some extent is comparable to its boundless crimes ... must lie heavily on the soul of those who were innocently swept up in it. I hope that you will gain confidence from the fact that your scientific and moral reputation has remained highly regarded all over the world, and everyone knows that you did not "howl with the wolves." ... Except for you and von Laue there is no one that enjoys such trust among the colleagues abroad, and it seems to me that is the most decisive factor for the reconstruction effort.... As you can see from this I am very interested in the future of German science and the KWG.[64]

Meyerhof continued that he wished "to express my sympathy and respect for all that you lived through and all the good you did or tried to do in the German hell. ... I hope that your son came back from the war and that your closest family is well." After saying he and his immediate family were safe, Meyerhof stated he had lost "countless relatives and close friends in the Polish gas chambers and concentration camps." Although Kuhn had asked him to resume his professorship in Heidelberg, he could not countenance a return to the country that had become "a graveyard for his nearest people." Meyerhof's poignant letter induced just one line of sympathy from Hahn, who understood "you are not inclined to return here anymore." Otherwise, Hahn expressed resentment for émigré German scientists, such as Meyerhof, who were prepared to criticize their countrymen "for the events of the last 12 years."[65]

Col. Blount's main concern in the summer of 1946 was to preserve the KWG, which the Americans, French, and Soviets wished to see disbanded because of its recent Nazi connections.[66] He outwitted the other powers in the Allied Control Council to keep it in existence. In September, he managed to secure its future, at least in the British zone, by renaming it the Max Planck Society (MPS). When he informed Hahn of the change in title, he threatened to resign, saying he had been appointed president of KWG and not of another society. This was an empty threat, as was his prediction that German scientists would now abandon the West in favor of Russia.[67] Hahn's inclination to blame others for the desperate condition of Germany was challenged by Meyerhof, when he wrote again in November:

> Germany is not only defeated, that is the least of it, it has gone through a moral catastrophe without precedent in history. I consider it completely essential for the recovery of intellectual life in Germany that this be made entirely clear to everyone, so that students will not continue growing up in the same morass.... What I want from you and all resolute non-Nazis is that you work to change this situation, even during the current state of hunger, in return for the work we are doing here to promote sensible politics with respect to German science.
>
> Now I must openly say, that even the best and most tried and true of my friends over there, such as yourself, have so far not been able to free themselves from the constricted perspective created by the Nazis, so as to understand the true roots of the resultant situation in Germany and Europe. ... My interpretation is that what Germany is now experiencing it entirely "owes to its Fuhrer" and it owes its Fuhrer to itself.[68]

Meyerhof's words did not lead to any soul-searching on Hahn's part. In a speech at the Nobel banquet in December, Hahn referred to Germany as "probably the most unfortunate country in the world."[69] In April 1947, Meyerhof summarized his own opinion concisely:

> Forgiveness is not my responsibility, but that of a higher power. Our duty is to keep the memory of this horror alive and thus to prevent future generations from living through it again. That is, I think the most important responsibility of German researchers and educators today.[70]

While Meyerhof was in support of the Max Planck Society rising from the ashes of the KWG and was even reconciled to the idea of Hahn as its president, he thought "they made the terrible mistake of taking over the Nazi administrator"[71] who became the secretary general of KWG in 1937. "This Nazi," he reminded Hill, was Ernst Telschow, the man who formally terminated Meyerhof's appointment at the Heidelberg KWG institute in 1938, and then thwarted an agreement Meyerhof made for a severance payment with the KWG. In Meyerhof's view, Telschow usurped the position of KWG president and imposed the system of *Gleichschaltung* or institutional standardization as required by the Reich. "From my own dismissal," he continued, "I know he was double talking and had the usual attitude of the ordinary Nazi liars." He was going to write to Hahn objecting to Telschow's continued employment and thought Hill might wish to inform the British Research Control. Hill did pass on the concerns to Col. Blount, who responded that Telschow had already been through the British and German denazification procedures and, despite there being no lack of accusers, he had been acquitted.[72] Meyerhof apologized to Hill[73] for taking his time and was happy accept Col. Blount's opinion, especially as Hahn had also written defending Telschow. Perhaps as a result of these assurances, when Hahn offered Meyerhof the honorary position of "External Scientific Member" of the MPS, he wrote a charming letter of acceptance:

> I would like to say that I consider the fourteen years' work as scientific member and director of the Institute of Physiology of the Kaiser Wilhelm Society in Dahlem and in Heidelberg as the most successful period of my scientific career, and that I have the fondest memories of these years and of the collaboration with my colleagues there, and that I am happy to resume these contacts.[74]

Time has shown Meyerhof's raw assessment of Telschow to be correct.[75]

From 1942, Dale combined the office of PRS with being director of the Davy-Faraday Laboratory at the Royal Institution (RI). If anything, the internal politics there were even more contentious at the RI than at Burlington House, the home of the RSoc. Dale wrote to Hill in March 1945 that the secretary of the RI refused to resign "although he has for years been unfit to do anything"[76] and the managers could not agree among themselves. His solution was for Hill to establish a laboratory for biophysics and then succeed him as director. Tizard's advice to Hill was unenthusiastic: "Don't you go to the RI unless you get a clear welcome and an unchallenged authority. UCL will only be too glad to have you back."[77] Dale succeeded in pushing the managers to offer the directorship to Hill and "all lab space not required by existing and statutory commitments"[78] for his new Biophysics Department. After close consideration, Hill thanked Dale and the managers and said he had decided to return to UCL.

The physiology laboratories at UCL had not been used for six years and were occupied by the Admiralty and Ministry of Food workers. His lab manager, Mr. Parkinson, returned from war service at RAF Farnborough full of enthusiasm, and he set about reclaiming territory and repainting walls. Equipment was both dusty and scarce. Parkinson acquired surplus items from Farnborough, and when Maurice Hill returned from a navy trip to Kiel with electrical apparatus for his father, he was given a shopping list for more to acquire on future visits to Germany such as a Siemens analyzer and a magnetic tape recorder.[79] Hill's standing in military circles was such that he had no hesitation in tackling the First Lord of the Admiralty, A.V. Alexander, when the building was still largely occupied by his people in 1946. In no uncertain terms, Hill complained that their presence was preventing a full intake of medical students and stopping admission of dental and science students all together. The electrical disturbance from Admiralty work was so strong that he was unable to carry on his research in the Biophysics Lab on the floor above them.[80]

Bernard Katz, who had worked in UCL and lived at Hurstbourne before the war, accepted a position in Australia in 1939 to work with John Eccles, a young neurophysiologist. When the war came, Katz joined the RAAF and took charge of a team responsible for mobile radar units both in Australia and New Guinea. Hill rated Katz as one of the most talented scientists he had ever encountered, and decided he wanted him as his deputy in the Biophysics Department. He wrote a report to the RSoc, "The need for an Assistant Director of the Biophysics Lab at U.C."[81] and said that Katz's name "jumped out." If funding were not found, he would have to resign as their foreign secretary. The threat worked instantly and the appointment confirmed one month later by a telegram from Sydney: "Rita and I very thrilled at wedding present will come soon as possible Katz."[82] On arrival in London, Bernard and his Australian wife moved into the top floor of Hurstbourne for the next three years.

Katz was an example of a physiologist already carrying out research on electrically excitable tissue (in his case, muscle and nerve), whose familiarity and dexterity with electronics were hugely advanced by his war service. Other illustrious examples of this breed were Alan Hodgkin, Andrew Huxley, and Richard Keynes, who formed the core of the "nerve team" in the Cambridge physiology department. Their wartime experiences as members of Blackett's circus or working directly in the services had indeed made them understand the advantages of a team approach, as well as encouraging quantitative and diagrammatic habits.[83] Katz became the London branch of the Cambridge nerve team and would meet the others in the summer at the MBS, often staying at Three Corners with Hodgkin and others.

There was another cohort of mostly younger physicists who had gained radar experience in the war and now wanted to start careers in biological sciences. Early in 1946, Hill placed an advert for such men to come to UCL for biophysics training and recruited two friends from the Telecommunications Research Establishment at Malvern, Eric Denton and J. Murdoch Ritchie.[84] Hill sent them both off to the MBS to soak in the basics of biology. In May 1947, he received a note from a slightly older physicist, who explained that he felt "a strong, though uninformed inclination to some form of bio-physics."[85] Hill invited him to attend a lecture at UCL and to meet him afterward. His name was Francis Crick, and Hill saw something in him. In 1939, Crick was a desultory doctoral student in Andrade's physics department but was selected from the RSoc's Central Register to work on mines for the Royal Navy. Hill told Crick that he needed to learn some basic biology and also arranged for him to see a Cambridge professor, who had an opening for a researcher to study color vision. He received a second letter from Crick in July saying that he had decided his interest was in the biophysics of the individual cell and not the special senses. Hill phoned Sir Eric Mellanby at the MRC and told him about Crick. He then told Crick that Mellanby agreed he needed a grounding in biology, but he should phone him at the MRC to set up an interview.[86] A few weeks later, Crick had an MRC grant for 400 pounds per year and Honor Fell as a supervisor. "As this was almost entirely due to your recommendation," he wrote to Hill, "I should like to express my very grateful thanks."[87] Hill was delighted and responded immediately:

> You might get some useful advice from the various young physiologists in the Physiology Department, my son D.K. Hill, Alan Hodgkin and Andrew Huxley who are all physicists by nature though with biological experience and knowledge. All have been working for the services themselves they will understand the position you are in. Go and talk to my son if you want an introduction. You will find him either there or Trinity College.[88]

Hill's final experiments during the phony war of 1939 were carried out in his garden shed on the mechanical efficiency of isolated frog muscle. If he was keeping up with the journals, he might have seen a short paper in *Nature* that October from two biochemists in Moscow. Engelhardt and Lyubimova, a husband-and-wife team at the Academy of Sciences, reported that the major structural protein of muscle, "myosin," had the ability to split the high-energy phosphate molecule, ATP.[89] They could not separate the enzyme responsible, adenosinetriphosphatase or ATPase, from the protein. One scientist who immediately took note of the paper from Moscow was Albert Szent-Györgyi, professor of medical chemistry at the University of Szeged in southern Hungary (Figure 12.2). He resolved: "It is evident, then, that we need to understand myosin if we want to understand contraction."[90] Szent-Györgyi had already won a Nobel Prize in 1937 for work on vitamin C and on the respiratory processes in muscle cells, research that laid the groundwork for Hans Krebs's elucidation of the Krebs cycle.

Working with younger members of his laboratory in Szeged in 1941, Szent-Györgyi was extracting "myosin" from freshly ground rabbit muscle using a standard technique that involved soaking in cold salt solution for 20 minutes, followed by spinning

Figure 12.2 Albert Szent-Györgyi (1893–1986)
Portrait of Albert Szent-Györgyi by unknown artist.
© The Royal Society.

in a centrifuge. One evening, the laboratory work was interrupted, and the mixture was left, unspun, in a fridge. The next morning, Ilona Banga found the preparation had turned to jelly overnight, much too viscous to separate by centrifuging. Although its total protein content was not much greater than a regular preparation of "myosin" extracted after 20 minutes, Szent-Györgyi postulated that the longer extraction process yielded a second protein. Szent-Györgyi called the viscous jelly "myosin B" and the 20-minute extraction "myosin A." Banga and her professor added a tiny quantity of ATP to myosin B, and its viscosity was drastically reduced, whereas it made no difference to myosin A. Another worker in the lab, Brúnó Traub, identified the new protein: since it altered the action of myosin it was named "actin." Myosin B was a complex of the two proteins, actomyosin. Szent-Györgyi went on to show that in the presence of ATP, potassium and magnesium, a thread of actomyosin underwent violent contraction. He was a scientist of imagination and passion, who later described witnessing the threads shortening, "to have reproduced in vitro one of the oldest signs of life, motion, was perhaps the most thrilling moment of my life."

Hungary was part of the Axis, and Szent-Györgyi was soon on the run from the Gestapo so that news of his startling discoveries did not reach the West until the end of the war. He came to London in the summer of 1946 and gave a lecture about the biochemistry of muscle contraction at the RI. Dale wrote excitedly to Hill about it:

His story of the two muscle proteins—myosin and actin—which cannot be stimulated separately, but which, when mixed in the appropriate proportion and in a balanced solution of inorganic ions, give a gel which contracts with great rapidity in response to adenosine triphosphate, is very fascinating. ... Doubtless you and Katz will be seeing him shortly.[91]

Dale was so impressed that he suggested that Szent-Györgyi should be invited to deliver the prestigious Croonian Lecture at the RSoc. When Szent-Györgyi visited UCL a few days later, Hill informed Dale that he gave him:

a number of elementary facts about muscle which he will have to reconcile with his observations on muscle fibres. For example, he found by his thermo-dynamic calculations ... that the heat of shortening of his fibres in ATP is negative. There is no doubt at all that when a muscle contracts the heat of shortening is pretty large and positive, that is the heat specially associated with the shortening process, over and above that resulting anyhow from stimulation and over and above the work.[92]

Coming from the man who had spent two decades repeatedly measuring the heat associated with muscle fiber activity, these were weighty criticisms. Szent-Györgyi promised to send him the wartime papers from Szeged which were written in English. In the meantime, Hill did not doubt his observations, but thought some of his explanations were "more poetic than scientific." Until he had the chance to study the work more critically, Hill cautioned against proposing him for the Croonian Lecture.

Szent-Györgyi also sent the original papers from Szeged to Meyerhof, who wrote to Hill that he was "very favourably impressed" by them since they "convey the feeling that he has made some important new observations, irrespectively of the phantastic and somewhat crazy ideas which he extracts from them."[93] He also mentioned that Szent-Györgyi had recently returned to Budapest after waiting in Paris for a promised US visa that never materialized. Szent-Györgyi was able to travel to the United States in the spring of 1947 and, as Meyerhof told Hill, "[he] has all people stirred and startled by the abundance of his discoveries and his imagination."[94] Meyerhof regretted that he would not be able to attend the first postwar International Congress to be held that summer in Oxford. When Hill replied, he was expecting some of Szent-Györgyi's "stuff" to be "pretty heavily criticized by those who know such things about myosin" in Oxford.[95] In fact it was Hill who challenged Szent-Györgyi at Oxford, when he was talking about the temperature dependence of actin-myosin interaction. Hill pointed out that his findings would lead to the conclusion that frogs would not swim at zero degrees, but they do. John Gergety, who had been a student in Budapest during the war, was there and explained many years later that "there was some little glitch in theory and we straightened it out ... the whole myosin field in the early days was based on a lot of assumptions."[96]

After the Oxford congress, Szent-Györgyi emigrated to the United States and settled in at the Woods Hole MBL, a place he fell in love with during the 1929 International Congress. He gave a careful demonstration of his crucial muscle experiment to Meyerhof the next summer, which he described to Hill:

> [A] muscle washed out with distilled H_2O and kept frozen for several months in glycerol, which still shows striation; washed out again and teased to bundles of very few fibers, and then put into Ringer's solution. Then a drop of dilute ATP is added and the bundle of fibers shrinks to a quarter and less of its length. If put on a tension lever, it develops tension which roughly calculated for one square centimeter would be far over a kilogram. Whatever the theories may be this astonishing experiment cannot be overlooked. He thinks it is quite specific for ATP.[97]

The methodology described here was devised by Szent-Györgyi to meet the criticism that the contraction of threads of actomyosin was not the same as in muscle fibers themselves. The preparation in glycerol allowed the separation of soluble muscle proteins from their membranes and "still showing striation" confirmed the characteristic microscopic appearance of intact skeletal muscle. The process also removed all traces of ATP from the protein, yet when a drop was added, the fibers contracted immediately.

Hill was still unconvinced that ATP provided the crucial energizing link between protein structure and true muscle contraction (which he did not accept Szent-Györgyi had achieved despite the striations). He wrote a letter to *Nature* challenging biochemists to come up with direct evidence:

> It is commonly stated on evidence obtained with muscle extracts, that the energy of muscular contraction is derived in the first instance from the breaking of the terminal energy-rich phosphate bond of adenosine triphosphate. Why not try to find out whether it really is, not in muscle extracts which cannot contract but in muscles which can?[98]

Meyerhof responded laconically three months later: "I do not know whether Szent-Györgyi's experiments are now more appreciated in England than they were formerly."[99] Meyerhof enjoyed a number of discussions with Szent-Györgyi at Woods Hole over the summer of 1949, including one he described to Hill in "the completely crowded lecture theater." At that seminar, Meyerhof argued that one should not accept

> the thermodynamic nonsense and arbitrary use of words like reversibility, muscle contraction, endothermic et cetera. I especially upheld your [Hill's] experimental results and what follows from them. I think I impressed him as well as the audience that he would do better to leave thermodynamics alone and to restrict himself to the study of acto-myosin and ATP. From the experiments with washed out muscle, I must infer that he has isolated an important partial process in the contraction mechanism. But one must insist it is a partial process. If this is acknowledged, then there are no contradictions between his results and your interpretation.[100]

Hill's response, while still begrudging in tone, showed that he was actively thinking about the issues:

> I hope very much that Szent-Györgyi takes your advice and sticks to things he understands. He undoubtedly has a flair for making curious biochemical discoveries but he writes the most appalling nonsense. . . . I should like to know whether the glycerol muscles relax as well as contract and whether the speed of shortening depends on the load in the way that it does in living muscles.[101]

The question about relaxation was deceptively profound: it was generally assumed that there was an, as yet undiscovered, inhibition mechanism or chemical relaxation factor that allowed muscle contraction to be reversed. Meyerhof answered Hill: "They do not relax, neither when ATP is present nor when it is washed out. Szent-Györgyi speaks of reversibility but this is thoroughly misleading."[102] It does appear that Hill's antipathy toward Szent-Györgyi was more than academic. Nobel Laureate George Wald (who worked in Meyerhof's laboratory for a short period) described a testy encounter at a banquet: "I sat beside a very famous muscle physiologist who when Albert's name came up somehow, turned a little red and said vehemently 'He's a charlatan!' "[103]

The centuries-old mystery of muscle movement was essentially solved by two consecutive papers in *Nature* in May 1954. By using X-ray diffraction and electron microscopy to reveal the structure of muscle fibers in unprecedented detail, it was established that contraction and relaxation are the result of constituent filaments sliding past each other. There is no actual shrinkage nor extension of protein molecules. One of the papers was submitted from the physiology department in Cambridge by Andrew Huxley and a German student, Rolf Niedergerke. Two decades later, Huxley reviewed the progress made and paid detailed attention to "the numerous contributions of A.V. Hill and his school."[104] He also noted that the challenge that Hill issued to the biochemists in 1949 was not answered until 1962 when "direct evidence of ATP utilization in excitable intact muscle" was finally proved.

Notes

1. A.V. Hill, "Science and Defence: Anglo-American Partnership in Research," *The Times* (17/6/41) reproduced in Hill (1960), 280–284.
2. Hill (1960), 280–284.
3. Crowther (1970), 232–233.
4. A.V. Hill, "The Use and Misuse of Science in Government" (1941), in Hill (1960), 57–66.
5. A.V. Hill (7/4/41) BBC broadcast to Germany AVHL I 2/9.
6. A.V. Hill A message from another world M&R: 554–555 AVHL I 5/6.
7. The United Nations charter was signed by the United States, United Kingdom, Soviet Union, and China on 1/1/42, with twenty-two other nations signing the next day.
8. https://api.parliament.uk/historic-hansard/commons/1942/dec/17/united-nations-declaration.
9. H. Channon "CHIPS," *The Diaries of Sir Henry Channon* (London: Weidenfeld, 1993), 347.
10. https://api.parliament.uk/historic-hansard/commons/1943/jan/21/jewish-and-other-refugees-visas.
11. https://api.parliament.uk/historic-hansard/commons/1943/mar/11/business-of-the-house.
12. https://api.parliament.uk/historic-hansard/commons/1943/may/19/refugee-problem.
13. https://api.parliament.uk/historic-hansard/commons/1943/may/19/refugee-problem.
14. O.F. Meyerhof to A.V. Hill (5/12/42) and Hill's reply (1/2/43) Royal Society Archive MDA/A/26/11/69 and MDA/A/26/11/68.
15. A.B. Ramsay to A.V. Hill (2/5/43) AVHL II 4/66.
16. A.V. Hill to A.C.G. Egerton (29/6/43) AVHL II 4/66.
17. H.H. Dale to A.V. Hill (23/7/43) AVHL II 4/66.
18. A.V. Hill to H.H. Dale (26/7/43) AVHL II 4/66.
19. A.B. Ramsay to A.V. Hill (27/7/43) AVHL II 4/66.
20. api.parliament.uk/historic-hansard/commons/1942/feb/05/scientific-and-industrial-research#S5CV0377P0_19420205_HOC_26.

21. Hill (1960), 312–313.

22. L.S. Amery to A.V. Hill (23/7/43) HD/6/8/6/5/130 R Soc Collections.

23. L.S. Amery to A.V. Hill (24/9/43) HD/6/8/6/5/137 R Soc Collections.

24. W.S. Churchill to A.V. Hill (30/10/43) HD/6/8/7/9/155 R Soc Collections. Roberts (2018) notes that "the concept of perverted science was a potent one" for Churchill, which he first expressed in a speech during the election of 1935 (393).

25. A.V. Hill, "Scientific Research in India, a Report to the Government of India" (London: Royal Society, 1945).

26. The chemist, Sir Shanti Bhatnagar, the physicists, Sir Venkata Raman and Homi Bhabha, and the botanist, Birbal Sahni.

27. A.V. Hill (undated), "Large Streams from Little Fountains Flow, Tall Oaks from Little Acorns Grow" M&R, 631–642 AVHL I 5/6.

28. Hill (undated), 631–642

29. Report of the Inter-Departmental Committee on Medical Schools (Goodenough Report) (London: HMSO, 1944). Among the report's recommendations were a restructuring of medical education to include more preventative medicine, co-education in medical schools and sex equality in hospital appointments, comprehensive training for specialists, and linking of all major hospitals to undergraduate teaching centers. All this would need much greater investment from the Exchequer.

30. Hill (1945).

31. J.N. Sinha, "British Scientific Opinion and India's Development: A Case Study of A.V. Hill," *Proceedings of the Indian History Congress* 70 (2009–2010): 404–412.

32. Memorandum from India Office to A.V. Hill (28/2/44) HD/6/8/7/10/18 Royal Society.

33. A.V. Hill to J.B.S. Haldane (2/4/43) MDA/A/25/16/8 Royal Society.

34. A.C.G. Egerton to A.V. Hill (5/12/43) AVHL II 5/42. The first two women, Kathleen Lonsdale and Marjory Stephenson, were elected in 1945. The first woman supported by Hill was Dorothy Needham, the muscle biochemist, in 1948. Honor Fell was elected in 1952, supported by Haldane and Hill.

35. A.V. Hill to M.N. Hill quoted in Angier 2 (1978):155.

36. https://api.parliament.uk/historic-hansard/commons/1944/jul/28/india#S5CV040 2P0_19440728_HOC_55.

37. Post-War Needs in Biology Committee (2/5/44) CMB/57/5 R. Soc. This broad initiative followed a letter to the secretaries from Blackett and Fowler in October 1943, concerned about the future provisions for physics research.

38. A.V. Hill to H.H. Dale (19/7/44) HD/6/2/10/10 R Soc.

39. A.V. Hill (7/12/44) . The need for an Institute of Biophysics HD/6/8/6/5/158 R Soc.

40. A.V. Hill and H. Munro Fox, "The Needs of Special Subjects in the Balanced Development of Science in the United Kingdom," *Notes and Records of the Royal Society of London* 4, no. 2 (1946): 133–145.

41. H. Herxheimer to A.V. Hill (29/5/44) MDA/A/28/11/38 Royal Society.

42. A.V. Hill to H. Herxheimer (31/5/44) MDA/A/28/11/37 Royal Society.

43. A.V. Hill to H. Blaschko (13/6/44) MDA/A/24//9/18 Royal Society.

44. O.F. Meyerhof to A.V. Hill (20/7/44) AVHL I 3/58.

45. W. Meyerhof (2002), 73–74.

46. O.F. Meyerhof to A.V. Hill (26/9/45) AVHL I 3/58.

47. Remy (2002), 118–240.

48. O.F. Meyerhof to A.V. Hill (26/9/45) AVHL I 3/58.
49. A.V. Hill to O.F. Meyerhof (10/10/45) AVHL I 3/58. The German Chemical Society stopped awarding its Richard Kuhn Medal in 2005 because of his support of the Nazi regime before and during World War II. Their assessment of Kuhn was as equivocal as Hill's sixty years before. Kuhn was not a NSDAP member.
50. R. Kuhn to O.F. Meyerhof (17/10/45) MPT50 M613/1/27.
51. O.F. Meyerhof to R. Kuhn (1/11/45) MPT50 M613/1/27. Deichmann (2002) summarizes Kuhn's work on nerve gases and the possible antidotes in Heidelberg and his denial about it after the war. All poison gas research gives rise to opprobrium but was widely practiced in both world wars, for example by Hill's dear friend Joseph Barcroft at Porton Down with Hill's active support in 1941, when he tried to obtain some hounds for Barcroft's experiments (see AVHL I 2/2).
52. W. Mashler to O.M. Meyerhof (9/10/45) MPT50 M613/1/16.
53. O.F. Meyerhof to A.E. Perlow (29/1/47) MPT50 M613/1/26.
54. A.V. Hill to O.F. Meyerhof (9/8/45) MDA/A/26/11/63 R Soc.
55. P. Collins, "Presidential Politics: The Controversial Election of 1945," *Notes and Records: The Royal Society* 65 (2011): 325–342.
56. E.N.daC. Andrade to C.S. Sherrington (14/2/45) quoted in Collins (2011).
57. A.V. Hill to H.T. Tizard (7/6/45) quoted in Collins (2011).
58. H.H. Dale to A.V. Hill (6/6/45) HD/6/8/7/1 RSoc.
59. A.V. Hill to E.N.daC. Andrade (10/6/45) AVHL II 4/3.
60. Collins (2011).
61. A.V. Hill, "Titles and Honours" and "The Secretaries of the Royal Society" (both undated) M&R: 266–272 AVHL I 5/5.
62. H.H. Dale to C. Attlee (27/10/45) HD/6/8/6/5/204.
63. A.V. Hill to O.F. Meyerhof (1/7/46) MPT50 M613 U Penn archives.
64. O.F. Meyerhof to O. Hahn (26/6/46) quoted in R.L. Sime, "The Politics of Memory: Otto Hahn and the Third Reich," *Physics in Perspective* 8 (2006): 3–51.
65. O. Hahn to O.F. Meyerhof (5/8/46) quoted in Sime (2006).
66. www.mpg.de/11957784/70-years-Max-Planck-Society (accessed 2/2/21).
67. M. Walker, "Otto Hahn: Responsibility and Repression," *Physics in Perspective* 8 (2006): 116–148.
68. O.F. Meyerhof to O. Hahn (8/11/46) quoted in R.L. Sime, "No Return: Jewish Emigres and German Scientists after the Second World War," in Charpa and Deichmann (2007), 245–262.
69. Quoted in R.L. Sime, *Lise Meitner: A Life in Physics* (Berkeley: University of California Press, 1996), 341.
70. Deichmann (2002).
71. O.F. Meyerhof to A.V. Hill (12/4/48) AVHL I 3/58.
72. A.V. Hill to O.F. Meyerhof (27/4/48) AVHL I 3/58.
73. O.F. Meyerhof to A.V. Hill (24/7/48) AVHL I 3/58.
74. O.F. Meyerhof to O. Hahn (16/10/48) quoted in Hofmann (2012).
75. Ernst Telschow (1889–1988) benefited in 1945 from Max Planck's attesting to his non-involvement in political circles and from Hahn's support. In fact, Telschow joined the NSDAP in May 1933 and eagerly emmeshed KWG projects into the Reich machinery. When fresh allegations against him were considered by the Senate of MPS in 1949, Hahn

threatened to resign because Telschow was irreplaceable. The brazen Telschow remained secretary general until the early 1960s, appointed to federal agencies and in receipt of civic honors. See A. Przyrembel, "Friedrich Glum and Ernst Telschow" (2004), www. mpiwg-berlin.mpg.de/KWG/Ergebnisse/Ergebnisse20.pdf (accessed 21/4/21).

76. H.H. Dale to A.V. Hill (28/3/45) AVHL II 4/72.
77. H.T. Tizard to A.V. Hill (4/4/45) AVHL II 4/72.
78. R. Cory to A.V. Hill (27/4//45) AVHL II 4/72.
79. A.V. Hill to M.N. Hill (10/10/45) AVHL II 5/31.
80. A.V. Hill to A.V. Alexander (4/2/46) AVHL II 4/3.
81. A.V. Hill, "The Need for an Assistant Director of the Biophysics Lab at U.C." (5/9/45) AVHL II 4/46.
82. B. Katz to A.V. Hill (12/10/45) AVHL 4/46.
83. Stadler (2009).
84. Both men became FRSs. Sir Eric Denton (1923–2007) became the director of MBS Plymouth, while Murdoch Ritchie (1925–2008) emigrated to the United States where, for many years, he was the head of the pharmacology department at Yale.
85. F.H.C. Crick to A.V. Hill (1/5/47) AVHL II 4/18.
86. A.V. Hill to F.H.C. Crick (4/7/47) AVHL II 4/18.
87. F.H.C. Crick to A.V. Hill (31/7/47) AVHL II 4/18.
88. A.V. Hill to F.H.C. Crick (1/8/47) AVHL II 4/18.
89. W.A. Engelhardt and M.N. Ljubimova, "Myosine and Adenosinetriphosphotase," *Nature* 144 (1939): 668–669.
90. Quoted in B. Bugyi and M. Kellermeyer, "The Discovery of Actin," *Journal of Muscle Research and Cell Motility* 41, no. 1 (2020): 3–9. Other sources for Szent-Györgyi: Rall (2014), 11–18; Needham (1971),146–68 and the Albert Szent-Györgyi Oral History Collection at https://www.nlm.nih.gov/hmd/manuscripts/oralhistory/gyorgyicoll.html.
91. H.H. Dale to A.V. Hill (26/7/46) AVHL I 3/12.
92. A.V. Hill to H.H. Dale (2/8/46) AVHL I 3/12.
93. O.F. Meyerhof to A.V. Hill (28/12/46) AVHL I 3/58.
94. O.F. Meyerhof to A.V. Hill (18/4/47) [as a p.s. to a letter written to AV by Gottfried Meyerhof] AVHL I 3/58.
95. A.V. Hill to O.F. Meyerhof (1/7/46) AVHL I 3/58. He also described Princess Elizabeth being admitted as an FRS, when she "behaved very nicely … neither bored nor blasé, nor shy."
96. oculus.nlm.nih.gov/cgi/t/text/text-idx?c=gyorgioh;cc=gyorgioh;rgn=main;view=text;idno=101318026-01.
97. O.F. Meyerhof to A.V. Hill (24/8/48) AVHL I 3/58.
98. A.V. Hill, "Adenosine Triphosphate and Muscular Contraction," *Nature* 163 (1949): 320.
99. O.F. Meyerhof to A.V. Hill (19/4/49) AVHL I 3/58.
100. O.F. Meyerhof to A.V. Hill (20/9/49) AVHL I 3/58.
101. A.V. Hill to O.F. Meyerhof (10/10/49) AVHL I 3/58.
102. O.F. Meyerhof to A.V. Hill (1/11/49) AVHL I 3/58.
103. Quoted by J.A. Rall, "Generation of Life in a Test-Tube: Albert Szent-Gyorgyi, Bruno Staub and the Discovery of Actin," *Advances in Physiology Education* 42 (2018): 277–288.
104. A.F. Huxley, "Muscle Contraction," *Journal of Physiology* 243 (1974): 1–43.

13
Legacies

The Marine Biology Lab (MBL) at Woods Hole was the perfect place for Otto and Hedwig to enjoy their summers together. Hedwig had always been a talented artist—as a young mother in Berlin she studied drawing with Johannes Itten, who was associated with the Bauhaus movement. She started taking art lessons again in Philadelphia and loved painting on Cape Cod. Her son, Walter, said she could communicate the shape and shadows of a sailboat on the water "with a single stroke of the crayon."[1] For Otto, it was a return to the familiar ambience of a laboratory by the sea, like the one in Naples where he carried out his first research. These laboratories began to appear in the second half of the nineteenth century and proved their worth in the first half of the twentieth, both for general marine studies and particularly in the areas of embryology, evolution, and neurophysiology. They allowed scientists from different backgrounds and countries to mix and exchange ideas over a period of weeks and were ideal settings for research and learning.

While Meyerhof enjoyed the coastal setting of Woods Hole, he valued it most for the intellectual exchanges with other scientists such as Szent-Györgyi. In the summer of 1949, he also had the opportunity to spend time there with his original, irascible mentor, Otto Warburg. In June 1948, Warburg came to the University of Illinois, Urbana, at the invitation of Bob Emerson, a botanist who had undertaken post-doctoral research with him in Berlin before the war. Emerson imagined they could collaborate on photosynthesis experiments to settle a difference of opinion over how many quanta of light are required for algae to produce one molecule of oxygen from one molecule of carbon dioxide. Warburg had been a leading figure in photosynthesis research since the early 1920s: he estimated that 4 quanta were sufficient, whereas Emerson and his colleagues determined, by experiment, the minimum to be 8–12 quanta. James Franck, the physicist, wrote to Meyerhof after a discussion at AAAS: "all agree that around 9 or 10 quanta are needed and not 4. ... Believe me, even the great Warburg can err sometimes."[2]

Warburg arrived in the country by plane with his valet and half-a-dozen large cases of belongings; he made it clear from the outset that he could not be expected to make any adaptations to his new surroundings (soon labeling his hosts the "Mid-West Gang"). As Hill presciently wrote to Meyerhof, Warburg "does not make things any easier for those who are ready to help him. ... I am afraid he is an awkward

Bound by Muscle. Andrew Brown, Oxford University Press. © Oxford University Press 2023.
DOI: 10.1093/oso/9780197582633.003.0013

customer."[3] Meyerhof wrote to Warburg regretting that he had not heard from him for ten years, which he assumed initially was due to "the political situation" but "after the German collapse, where there is no longer any danger for you, I could have received a sign of life from you."[4] Meyerhof's letter reached Warburg just as his time in Urbana was drawing to a bad-tempered climax. Emerson organized a seminar before Christmas to try, finally, to entice Warburg into joint experiments, but his efforts were in vain and Warburg left in anger, without saying goodbye.[5]

Warburg's infamous capacity for personal grudges was unleashed:

> Lieber Meyerhof,
>
> To have it out:
>
> 1/ When you left Germany in 1938, you wrote me a letter that began: "Now that we are living in separate camps." Considering how much he had helped Meyerhof "from the first to the last day, I did not understand this sudden and unmotivated hostility."
>
> 2/ "After the collapse you probably knew how the people in Germany were doing. I was surprised you didn't try to help me."

Warburg closed by saying he had been in the United States for six months and while most people had been charming, "Emerson was abominable." Warburg survived the Nazi regime outwardly unscathed, treating it with open contempt. Even though he had two Jewish grandparents and was homosexual, he was permitted to keep his laboratories operating throughout the war.[6] His fusillade against Meyerhof was unjustified and wounding. He wrote again in April trying to entice Meyerhof to join his institute in Berlin-Dahlem, and then in May announced he would be coming to Woods Hole in mid-June for a long stay and looked forward to seeing him.[7]

Warburg arrived with Dean Burk, a senior chemist who had been his host at the NCI for six months, and a copious amount of equipment. Burk was an enthusiastic promoter of Warburg's opinion on the high efficiency of photosynthesis, and the pair gave the impression that the argument was settled in their favor. The controversy bore some similarity to the one about dehydrogenation with Wieland thirty years earlier but was even more rancorous. Warburg refused to speak to Emerson and made no real effort to address the criticisms raised at a lively seminar at Woods Hole.[8] Warburg and Meyerhof, though, enjoyed each other's company that summer (Figure 13.1). David Nachmansohn, who was also in summer residence, described a conversation with Warburg after they had spent an evening sitting in Meyerhof's garden:

> [Meyerhof] discussed modern trends in biology and some of the recent remarkable achievements, frequently adding some interesting and stimulating philosophical comments. Warburg was visibly fascinated. On the way home he told me, "You know he is the greatest personality of all of us."[9]

Figure 13.1 Scientists at Woods Hole, summer 1949. (left to right) Saul R. Korey, David Nachmansohn, Dean Burk, Albert Szent-Gyorgyi, Otto Warburg, Otto Meyerhof, Carl Neuberg, George Wald

Theodore Bullock Papers. SMC 52. Special Collections & Archives, UC San Diego Library.

Meyerhof wrote to Hill about his own impressions after Warburg returned to Berlin:

> After I made some conciliatory moves, Otto Warburg was very friendly and I saw a great deal of him in Woods Hole. All together he was more mellow and jovial as is his nature. … He is, of course, a very queer and intractable person but also highly stimulating. It depends on what one takes as more important.[10]

Such was Meyerhof's standing as a biological chemist that his colleagues and students from Europe and the United States published a Festschrift, *Metabolism and Function*, to celebrate his sixty-fifth birthday.[11] Nachmansohn was the organizing editor, and Hill wrote the introduction:

> The results of his researches, and those of his colleagues, are a part of scientific history. They are linked with most that is known of the chemistry of muscle and with much that is established of changes involving phosphate and carbohydrate in the cell. For some years his investigations were concerned mainly with muscle—living muscle: more recently they followed the trend in biochemistry, perhaps even they helped to establish it, of dealing *in vitro* with the enzyme systems of muscle.[12]

Strangely, Hill seemed to find it difficult to accept that his old friend had left the field of physiology, where some semblance of organ function remained central, for biochemistry, where molecules reign supreme. He went as far as repeating his challenge to biochemists to show that ATP is the immediate source of energy underpinning muscle contraction. Meyerhof was delighted with the Festschrift and when acknowledging Hill's contribution, added, "I hope the challenge is at least taken up by other biochemists if not by myself."[13] He mentioned in the same letter that Richard Kuhn, his antagonist from Heidelberg, now in comfortable possession of his Nobel Prize and still director of the Max Planck medical institute, was visiting the University of Pennsylvania. Meyerhof displayed remarkable personal forgiveness toward Kuhn: "I am very friendly to him in spite of former disagreements." In September 1950, Meyerhof wrote to Kuhn asking him to intercede with the Max Planck Society (MPS), which he believed should finally fulfill "its obligations towards the people who were dismissed under Nazi pressure instead [of] trying to placate them with promises for the future."[14]

Meyerhof used the occasion of the bicentenary of Goethe's birth to express his "ambiguous feelings" about his home country. Speaking in German to the Rudolf Virchow Medical Society in New York, he praised Goethe, whom he had adored since his teens, as Germany's greatest gift to the world, comparing his "vivid expression of all-embracing humanity" with "the Germany of today [that] has betrayed this heritage and violated it beyond any conceivable degree."[15] Paradoxically, the moral disparity made celebrating Goethe even easier because of "his cosmopolitanism ... the transnational, all-penetrating spirit of his being. Thus, we preserve the pure flame of his vivid spirit, even if all places of Goethe lie in ruins as a horrible symbol of this betrayal." The title of the address was "About Goethe's Method of Nature Research," and Nachmansohn was struck by the careful way Meyerhof took the side of scientific evidence over Goethe's descriptive contributions because:

> The scientific analysis of nature was not Goethe's real goal. It was the search for the deeper meaning of creation—*die Ahnung des Ewigen in Endlicken*—to use the words of Fries.[16]

In December 1950, Hill traveled to Stockholm to take part in celebrations for the fiftieth anniversary of the Nobel Prizes. As the senior laureate in physiology or medicine attending, he made a speech of thanks at the banquet in his name and Meyerhof's. He wrote an account to Meyerhof to tell him how many people had asked to be remembered to him. Meyerhof's poor health made it impossible for him to travel, and he also complained to Hill it would have cost more than $1,000 for him and Hedwig to attend—a foolish amount to spend. He mentioned that Warburg had also written to him about the occasion, before adding that Warburg was aware of Hill's "understandable disinclination" toward him. For his part, Meyerhof thought he understood "the queer working of his mind and am always impressed by his scientific greatness which has not lessened with the years [so] I try somehow to re-main above the situation."[17]

In June 1951, a symposium was held at Johns Hopkins on the role of phosphates in metabolism. Since so many of the distinguished participants were former and cur-rent colleagues of Meyerhof's from Germany and the United States, he was invited to give the opening paper.[18] He surveyed with pride the rapid progress in a subject that did not exist twenty-five years earlier. He started by quickly reviewing fermentation and glycolysis, subjects that had fascinated him since before World War I. Glycolysis begins with the phosphorylation of sugar, initiating a series of successive com-pounds, fabricated by different enzymes, that serve as the intermediate steps to the ultimate formation of pyruvate. This sequence produces a net gain of two molecules of ATP from each molecule of glucose. In the presence of oxygen, the pyruvate then enters the Krebs cycle, which in turn feeds the process of *oxidative phosphorylation* inside the mitochondria of cells, yielding a total of thirty-four molecules of ATP. The analysis of high-energy phosphate bonds in compounds such as ATP began in Meyerhof's Heidelberg laboratory; extended by his co-workers, notably Lohmann and Lipmann, by 1951 the knowledge gained, as Meyerhof stated, "has modified our whole thinking in various fields of physiology." The transient bonds are the thermo-dynamic currency of life.

Otto and Hedwig lived in Hamilton Court, an apartment complex close to the University of Pennsylvania. Richard Kuhn, visiting Philadelphia again, was also staying in Hamilton Court and spent two hours in conversation with Otto on Friday, 5 October 1951. Although the content of the conversation is not known, it seems likely that they would have spoken about the effort to recoup the compensation that Meyerhof was owed from his prewar dismissal as director of the KWG institute, a process that Kuhn had been prosecuting on his behalf and which Telschow had been obstructing. On the Saturday morning, Meyerhof, who had regarded himself as living on borrowed time since his heart attack in 1944, suffered a fatal myocardial infarc-tion.[19] Since he worked as a research professor only for a few years and received a stipend from the Rockefeller Foundation, Hedwig would receive only $15 per month from the University of Pennsylvania. Otto Hahn was informed and arranged for a widow's pension from the MPS.

Hedwig moved out of Philadelphia and, released from caring for Otto, she immersed herself in art. She became a qualified art teacher, but sadly her newfound career and life ended in an automobile accident in 1954. The Meyerhof children all led full lives. Gottfried, self-anglicized to Geoffrey, emigrated to Canada where he became a leading structural engineer and expert on soil mechanics. Bettina Emerson-Meyerhof was a popular pediatrician working among underserved communities in Seattle. Walter became a professor of physics and headed the physics department at Stanford University in the 1970s.

Meyerhof was regarded by his contemporaries as a decisive figure in the develop-ment of biological chemistry, especially for his groundbreaking work on glycolysis and cell energetics involving high-energy phosphate bonds. Younger colleagues were mainly attracted by his published work, and he assembled powerful teams in Dahlem and Heidelberg. At least three of them would go on to win Nobel Prizes themselves: Fritz Lipmann, Severo Ochoa, and George Wald. Despite his reserved manner, his colleagues came to value him for his personal integrity as well as his intellect. Under the dark clouds of the Nazi regime, he was one of the few remaining professors unafraid to keep to his moral principles and went out of his way to protect threatened members of staff, whether they were communists or Jews, laboratory scientists or technicians. With the unobstructed clarity of hindsight, it is impossible to understand why he stayed so long in Germany or why he did not seek the escape route offered by the SPSL, where his friend Hill was the dynamic leader.

In his 1939 Abraham Flexner Lecture, Szent-Györgyi made a humorous observation about biochemistry:[20]

> To the outside spectator, all this work of the biochemist in which he shifts little H atoms and the phosphate molecules from one substance to the other must seem a little [like] play for big children. Thank heaven that this is rightly so, and that biochemistry is a lovely game of refined cookery, very fit for the amusement of big children.

Unlike Szent-Györgyi, Meyerhof was never lighthearted in the laboratory: they both made fundamentally important discoveries about the biochemistry of muscle, and Meyerhof was a most skillful manipulator of organic molecules. As we have seen, Hill was skeptical about the Hungarian's work, and Szent-Györgyi had a dig at Hill's own research, which had become repetitive over decades. In the second edition of his monograph, *Chemistry of Muscle Contraction*, published in 1951 after he had experienced years of sniping from Hill, Szent-Györgyi wrote:

> To the physiologist muscle is a unit ... which he tries to study in undamaged muscle and under conditions in which Nature has intended them to work, measuring changes in shape, tension, potential, heat, optical properties etc. These have been extensively studied ... leaving behind an enormous bulk of literature and little understanding.[21]

It is hard to imagine any reader familiar with the field not thinking of Hill's work first. He continued to repeat over a period of roughly half a century, with ever greater precision, the heat production during various modes of muscle contraction and relaxation. The improved methods depended largely on improvements in apparatus such as galvanometers and recording methods, which largely depended on others. With the advances in biochemical understanding and the revolutionary sliding filament model of muscle action, there was really no scientific value to his laboratory activity. While visiting the United States in 1927, he had surprised himself by admitting that his research was not driven by chasing any big idea but was for his own amusement.

When one of his grandsons, Nick Humphrey, won a scholarship to Trinity College Cambridge, AV took him into his lab at UCL the day after Christmas, almost as a rite of passage.

> On Boxing Day, 1960, soon after breakfast, Gower Street in London was deserted. I and my grandfather, A.V. Hill, entered the Anatomy Department of University College through a side door and made our way stealthily upstairs to his laboratory. The atmosphere was morgue-like, and a musty smell of formaldehyde hung in the air. Water dripped from the lab ceiling and splashed onto an umbrella raised over the bench. A clock ticked, oddly out of tempo with the dripping; otherwise there was an eery stillness. Grandpa removed the lid from a basin filled with live frogs, picked one out, and eyed its strong thigh muscles. He put it aside in a glass jar and called me over to admire it. The dissecting instruments and pins were waiting beside the corkboard.[22]

Hill's style as a scientist was akin to Rutherford's school of physics. The generation of reliable experimental data was the most valued business of the day—fitting the observations into overarching theories could wait. Like Rutherford, Hill was blessed with an imagination that often allowed him to see underlying processes not obvious to others: for example, the way oxygen combines with hemoglobin; how the angular momentum of an artillery shell could determine whether it would explode; and how oxygen restored lactic acid to its precursor state after prolonged muscle activity in accordance with the laws of thermodynamics. Meyerhof, with his love of art and poetry, his deep interest in post-Kantian philosophy, and his fascination with the chemistry of life, was consciously an intellectual and regarded as such by his contemporaries. Hill was not and is more accurately cast as an English pragmatist. Polly Hill, who followed her father's interdisciplinary tendencies by combining economics and anthropology while studying cocoa farming in Ghana, believed that *Lorna Doone* was the only novel he ever read, although he was steeped in the literature of the Bible and Shakespeare from his schooldays. Polly also admitted she had to resort to various subterfuges "to conceal part of my overwhelming admiration for AV."[23]

Another legacy from Blundell's school was Hill's love of running—a habit he continued until his knees gave way in his seventies. Athletics undoubtedly led him into exercise physiology in the early 1920s where he was an influential pioneer, inventing the concept of maximal oxygen uptake, $\dot{V}O_{2max}$.[24] In the summer of 1951 he received a letter from Harold Abrahams, who won the gold medal for the 100 m sprint at the Paris Olympic Games in 1924, asking him for his thoughts on running in the wind. Hill's reply contained incidental advice for the most promising middle-distance runner in the country, who was coming ever closer to a four-minute mile:

> Referring to running at a constant speed, I am sure Bannister would do something very near 4 minutes to the mile were he to run at a constant speed, perhaps rather quicker in … say the first 300 yd. Quite properly he has been more

> concerned with winning races than with breaking records and there the technique is quite different. From the point of view, however, of breaking a record in a mile, to do the last quarter in 58 seconds proves quite clearly that you ought to have run faster at an earlier stage.[25]

Abrahams replied he would discuss Hill's recommendations with Roger Bannister, who at that time had no trainer. He mentioned Bannister was interested in physiology and about to start his clinical studies at St. Mary's Hospital in London: "I know that if you could ever spare the time, he would be delighted at the opportunity of meeting you."[26]

In addition to founding the applied science of exercise physiology, Hill was a seminal figure in the development of biophysics. He saw the potential of converting young men who had gained experience with electronic recording in World War II; his protégé, Bernard Katz, who lived with the Hills for years, shared the 1970 Nobel Prize for Physiology or Medicine for his work on neurotransmitters (acetylcholine) at the junction between nerves and muscles. Hill also patiently supported Francis Crick in the late 1940s, arranging his entré into biological research at Cambridge. Crick wrote to Hill after being there for about eighteen months, saying that he was going to try to join the group under Max Perutz at the Cavendish Laboratory studying protein structure, even though Hill had advised against it.

> I've had the opportunity recently to look more closely into X-ray analysis, and I don't find it nearly as difficult as I expected. Whether it will prove tedious remains to be seen, but with three of us, and may be more in the group ... [it] should go at a fair rate. Moreover, I think I have the sort of brain which enjoys puzzles of this sort.[27]

As ever, Hill was supportive, wishing Crick "all success":

> Thank you for letting me know about your decision. I expect you are quite right. If the X-ray diffraction patterns of protein are what interest you most, in spite of any deterrent I may have exerted, you can be reasonably sure that your decision is the best one. ... I seem to remember now, when Perutz asked me about possible people to join him, I mentioned your name. If so, I don't regret it in spite of what I said.[28]

Sounding a familiar theme, Hill hoped that Crick would focus on living muscle, unlike the biochemists who usually work on muscle "ground up with sand." Crick did indeed join Perutz's group and became quite a disruptive figure, causing the director of the Cavendish, W.L. Bragg, to write to his old friend in January 1952:

> Dear Hill,
> There is a young man working here, in Perutz's team, who I believe at one time was a protégé of yours and advised by you to take up biophysics. This

is <u>Crick</u>. I am worried about him and if you take more than a passing interest in him, I should like to consult you about him. He is working for a PhD here, though he is 35, because the war stopped him trying before. My worry is that it is almost impossible to get him to settle down to any steady job and I doubt whether he has got enough material for his PhD which should be taken this year. Yet he is determined to do nothing but research and is very keen to hang on here. With a wife and family he ought to be looking for a job. I think that he overrates his research ability, and that he ought not to count on getting a job with no other commitments. Are you interested in his career enough to wish to discuss it? I should like some help in deciding what line to take with him

Yours ever

W. L. Bragg[29]

Hill took the trouble to reassure Bragg, who was therefore saved from making perhaps the greatest managerial blunder in the history of science: Crick and James D. Watson's description of the double helix of DNA came almost exactly one year later.

There had been slow progress toward establishing biophysics as a university department at UCL, despite Hill's best efforts in the late 1940s. He engaged the support of Sir Henry Tizard, now chief scientific adviser to the Ministry of Defence, who was eventually able to inform Hill that "no less a person than the Chairman of the University Grants Committee [guarantees] that the survival of the Biophysics Lab is a certainty."[30] Hill relinquished his Foulerton research chair and became its first director for the last few months of 1951. He was elected president of the British Association for the Advancement of Science (BAAS) in succession to the Duke of Edinburgh. Early in 1952, Hill and Tizard made a visit to Pakistan to support the PAAS conference in Peshawar. Hill delivered his BAAS presidential address, "The Ethical Dilemma of Science,"[31] in Belfast in September. It was an unusually wide-ranging lecture, denounced in turn by *Pravda* and the Vatican, which reassured its author. He returned to familiar themes of public health, overpopulation, and the education of women, as well as the new development of nuclear weapons. Rejecting the notion of a "scientific age," he posed the following:

It is true that the external circumstances of life have been vastly altered by the application of scientific discovery and invention, though as yet for only a minority of mankind. The future alone can decide whether natural resources and human ingenuity will prove sufficient, given statesmanship and goodwill, for the same transformation gradually to affect the whole of human society. If not, are stable conditions ultimately possible? Or will there be perpetual conflict between the "haves and the "have-nots"?

His peroration combined genuine modesty and wisdom:

> Is there really then any special ethical dilemma which we scientific men, as distinct from other people, have to meet? I think not: unless it be to convince ourselves humbly that we are just like others in having moral issues to face. It is true that integrity of thought is the absolute condition of our work, and that judgments of value must never be allowed to deflect our judgments of fact. But in this we are not unique. It is true that scientific research has opened up the possibility of unprecedented good, or unlimited harm, for mankind; but the use that is made of it depends in the end on the moral judgments of the whole community of men. It is totally impossible now to reverse the process of discovery; it will certainly go on. To help to guide its use aright is not a scientific dilemma, but the honourable and compelling duty of good citizen.

In addition to his presidency of BAAS, Hill also became secretary general of the International Council of Scientific Unions for four years, during which time he traveled widely and persuaded the Soviet Academy to join. In 1955 he was elected president of the Marine Biological Association, which gave him good reason to visit Plymouth. He especially loved taking grandchildren with him, staying at Three Corners, and arranging for them to go out on boats from the MBS. AV and Margaret would hold regular Sunday tea parties at Hurstbourne where children, grandchildren, and friends would gather. One occasional guest was a teenage Stephen Hawking, who was staying with the Humphrey family while his parents were away in India.

AV and Margaret celebrated their Golden Wedding in June 1963 at the Cambridge home of Maurice and Philippa. It was a delightful occasion with all four children and fourteen grandchildren present. The previous year, Maurice, an oceanographer, been elected as an FRS, which delighted his parents. He developed a seismic method for studying the ocean floor, revealing for example that there is 8,000 ft. of sediment at the bottom of the English Channel above bedrock. However, his long voyages in the Indian Ocean and Atlantic had taken a psychological toll on him, and he started to drink heavily. Fearing that his mental capacity was failing, in January 1966 he committed suicide at home using a shotgun, while his children were away. Neither AV nor Margaret could understand or bear to talk about his death. Margaret was already becoming incapacitated by Parkinson's disease and it was decided that they would move into Maurice's home in Chaucer Road, Cambridge, to live with Philippa and her children.

That summer saw AV's eightieth birthday and the Physiology Society held a dinner for him. One of the guests, Wilhelm Feldberg, a German Jew who had come to England through the auspices of the AAC in 1933 and gone on to become "one of the greatest neuropharmacologists of the 20th century,"[32] decided it was time to honor Hill the humanitarian. At Hill's suggestion, the dinner was held at the English-Speaking Union, where, as Feldberg observed, the seventy guests set a record for the number of English dialects spoken. They included numerous Nobel laureates, dozens of university professors, journalists, artists, and historians. Hill in his speech paid special attention to Esther Simpson, who was there and had looked after the welfare of them all when they arrived three decades before.[33] He finished with two cautionary quotations from the New Testament on the danger of listening "when all men shall speak well of you for so did their fathers to the false prophets" and a warning against being spoilt "through philosophy and vain deceit." This might have amused Karl Popper.

Margaret died in 1970 and AV became increasingly incapacitated by weakness in the legs. His last project was to collect his non-scientific writings together as "Memories and Reflections." When he could not find a publisher, he gave it to the Churchill College Archives. They arranged for a small number of printings, one of which went to his nephew, Richard Keynes FRS, professor of physiology at Cambridge. He wrote to AV to say the reason he became a physiologist was "I knew you! And I am eternally grateful for it."[34] As Maurice's cousin and closest friend, he said how much he missed him—"Life always seemed twice as full when he was around." The Physiology Society organized a ninetieth birthday party for AV at Pembroke College, where he was delighted to see his oldest son, David (FRS, 1972), as well as Katz, Denton, Huxley, Hodgkin, Rushton, Feldberg, and Richard Keynes. He was presented with a red leather-bound volume of 150 letters from friends. Sir Roger Bannister wrote:

> Professor A.V. Hill was one of the first to see that the real purpose of athletics was not to select Olympic champions but was to provide the physiologists with a series of experiments in which the human body was placed under stress on the path towards a series of absolute breaking points!

When Hill decided to move to Manchester in 1920, he wrote a letter to his friend and former tutor, Walter Fletcher, saying that life was the great game and he had to play it.[35] He kept his word almost until the day he died on 3 June 1977.

Notes

1. W. Meyerhof (2002), 104.
2. J. Franck to O.F. Meyerhof (21/1/48) MPT50 M613/1/16.
3. A.V. Hill to O.F. Meyerhof (1/3/48) MPT50 M613.
4. O.F. Meyerhof to O.H. Warburg (9/12/48) and Warburg to Meyerhof (2/1/49) MPT50 M613/1/21.
5. E. Rabinowitch, "Robert Emerson (1903–1959)," www.nasonline.org/publications/biogr aphical-memoirs/memoir-pdfs/emerson-robert.pdf (accessed 18/3/21), and K. Nickelsen, "Of Light and Darkness: Modelling Photosynthesis, 1840–1960," Habilitation thesis (Bern) (2011) www.core.ac.uk/download/pdf/12174538.pdf (accessed 18/3/21).
6. Apple (2021).
7. O.H. Warburg to O.F. Meyerhof (27/4/49 and 6/5/49) MPT50 M613/1/21.
8. Nickelsen (2011).

9. Nachmansohn (1972).
10. O.F. Meyerhof to A.V. Hill (20/9/49) MPT50 M613.
11. D. Nachmansohn (ed.), *Metabolism and Function* (New York: Elsevier, 1950).
12. A.V. Hill, "A Challenge to Biochemists," in Nachmansohn (ed.) (1950), 4–11.
13. O.F. Meyerhof to A.V. Hill (16/12/49) AVHL I 3/5.
14. Hofmann et al. (2012).
15. O.F. Meyerhof, "Über Goethes Methode der Naturforschung," *Proceedings of the Rudolf Virchow Medical Society* 8 (1950): 3.
16. Nachmansohn et al. (1960).
17. O.F. Meyerhof to A.V. Hill (3/1/51) AVHL I 3/5.
18. O.F. Meyerhof, "Phosphorus Metabolism," *American Scientist* 39, no. 4 (1951): 682–687.
19. Hofmann et al. (2012).
20. Oral history on Szent-Györgyi, oculus.nlm.nih.gov/cgi/t/text/pageviewer-idx?c=gyor-gioh;cc=gyorgioh;rgn=full%20text;idno=101318026-02;didno=101318026-02;view=image;seq=5;page=root;size=l;frm=frameset.
21. Quoted by Rall (2018).
22. N. Humphrey, "A Family Affair," in J. Brockman (ed.), *Curious Minds: How a Child Becomes a Scientist* (New York: Pantheon Books, 2004), 3–12.
23. P. Hill (1966). Note on A.V. Hill AVHL II 4/18.
24. Bassett (2002).
25. A.V. Hill to H.M. Abrahams (27/7/51) AVHL II 4/1 Abrahams's triumph at the Paris Olympics was immortalized in the movie *Chariots of Fire.*
26. H.M. Abrahams to A.V. Hill (31/7/51) AVHL II 4/1. When Bannister became the first to run a mile in under four minutes in 1954, his early pacemakers played a crucial role.
27. F.H.C. Crick to A.V. Hill (7/3/49) AVHL II 4/18.
28. A.V. Hill to F.H.C. Crick (11/3/49) AVHL II 4/18.
29. W.L. Bragg to A.V. Hill (18/1/52) AVHL II 4/18.
30. H.T. Tizard to A.V. Hill (6/3/51) AVHL I 3/95.
31. Hill (1960), 72–89.
32. G. Bisset and T.V.P. Bliss, "Wilhelm Siegmund Feldburg (1900–93)," *Biographical Memoirs of Fellows of the Royal Society* 43(1997): 145–70.
33. A.V. Hill, "Retrospective Sympathetic Affection" M&R, 598–603 AVHL I 5/6.
34. R.D. Keynes to A.V. Hill (18/3/74) AVHL II 4/48.
35. A.V. Hill to W.M. Fletcher (17/1/20) AVHL II 8/1.

Selected Bibliography

Apple, S. *Ravenous: Otto Warburg, the Nazis, and the Search for the Cancer-Diet Connection.* New York: Liveright Publishing, 2021.

Aubin, D., and C. Goldstein (eds). *The War of Guns and Mathematics.* Providence, RI: American Mathematical Society, 2014.

Ball, P. *Serving the Reich.* London: Vintage, 2014.

Barclay, C.J., and N.A. Curtin. "The Legacy of A.V. Hill's Nobel Prize Winning Work on Muscle Energetics." *Journal of Physiology* 600 (2022): 1555–1578.

Bates, A.W.H. *Anti-Vivisection and the Profession of Medicine in Britain.* London: Palgrave-Macmillan, 2017.

Beveridge, W. *A Defence of Free Learning.* Oxford: Oxford University Press, 1959.

Brecht, B. *Refugee Conversations.* London: Methuen, 2019.

Brown, A.P. *The Neutron and the Bomb.* Oxford: Oxford University Press, 1997.

CARA [Council for Academics At Risk]. www.cara.ngo/who-we-are/our-history.

Charpa, U., and U. Deichmann (eds.). *Jews and Sciences in German Contexts.* Tübingen: Mohr Siebeck, 2007.

Clark, R.W. *The Rise of the Boffins.* London: Phoenix House, 1962.

Clark R.W. *Tizard.* Cambridge, MA: MIT Press, 1965.

Cooper, R.M. (ed). *Refugee Scholars: Conversation with Tess Simpson.* Leeds: Moorland, 1992.

Crowther, J.G. *Fifty Years with Science.* London: Barrie & Jenkins, 1970.

David T.R.V. "British Scientists and Soldiers in the First World War (with Special Reference to Ballistics and Chemical Warfare)." PhD diss., University of London, 2009.

Deichmann, U. "Chemists and Biochemists in the National Socialist Era." *Angewandte Chemie International Edition* 41 (2002): 1310–1328.

Edmonds, D. "Esther Simpson: The Unknown Heroine." 2017. www.thejc.com/news/features/esther-simpson-the-unknown-heroine-1.438317.

Emmerich, M. "Cells Flexing Their Muscles." MaxPlanckResearch Flashback Biochemistry, 2013. www.mpg.de/7023399/S005_Flashback_086-087.pdf.

Fenn, W.O. "A Quantitative Comparison between the Energy Liberated and the Work Performed by the Isolated Sartorius Muscle of the Frog." *Journal of Physiology* 58 (1923): 175–203.

Fenn, W.O. (ed.). *History of the International Congresses of Physiological Sciences 1889–1968.* n.p.: American Physiological Society, 1968, 1968.

Gilbert, Martin. *The First World War.* New York: Henry Holt, 1994.

Gowing, M. *Britain and Atomic Energy.* London: Macmillan, 1964.

Harden, A. *Alcoholic Fermentation.* London: Green & Co., 1923.

Hardy, G.H. *A Mathematician's Apology.* Cambridge: Cambridge University Press, 1967.

Heim, S., C. Sachse, and M. Walker (eds.). *The Kaiser Wilhelm Society under National Socialism.* Cambridge: Cambridge University Press, 2009.

Helmholtz, H. "On the Conservation of Force [1847]." In *Scientific Memoirs Selected from the Transactions of Foreign Academies of Science,* ed. J. Tyndall and W. Francis, 114–162. London: Taylor and Francis, 1853.

Hill, A.V. "Autobiographical Sketch." *Perspectives in Biology and Medicine* 14 (1970): 27–42. https://doi.org/10.1353/pbm.1970.0009.

Hill A.V. *The Ethical Dilemma of Science.* New York: Rockefeller Institute Press, 1960.

Hill A.V. "The Heat Production of Muscle and Nerve, 1848–1914." *Annual Review of Physiology* 21 (1959): 1–19.

Hill, A.V. *Living Machinery*. New York: Harcourt, Brace & Co., 1927.

Hill, A.V. "Nobel Lecture 1922." www.nobelprize.org/prizes/medicine/1922/hill/lecture/.

Hill, A.V. *Trails and Trials in Physiology*. Baltimore: Williams & Wilkins Co, 1965.

Hodgkin, A.L. "Beginning: Some Reminiscences of My Early Life." *Annual Review of Physiology* 45 (1983): 1–16.

Hofmann E., R. Ulrich-Hofmann, and W. Höhne. "Otto Meyerhof and the Exploration of Glycolysis—Outstanding Research in an Inhumane Era." *Acta Historica Leopoldina, Vorträge und Abhandlungen zur Wissenschaftsgeschichte* 59 (2012): 317–382.

Issenberg, S. *A Hero of Our Own: The Story of Varian Fry*. New York: Random House, 2001.

Katz, B. "Archibald Vivian Hill (26/9/1886–3/6/1977)." *Biographical Memoirs of Fellows of the Royal Society* 24 (1978): 71–149.

Langen, P., and F. Hugo. "Karl Lohmann and the Discovery of ATP." *Angewandte Chemie International Edition* 47 (2008): 1824–1827. www.angewandte.org.

Lanouette, W. *Genius in the Shadows*. Chicago: University of Chicago, 1992.

Leal, F. "Who Was Leonard Nelson?" In Leonard Nelson, *A Theory of Philosophical Fallacies*, 6–9. Heidelberg: Springer, 2016.

Longair, M. *Maxwell's Enduring Legacy: A Scientific History of the Cavendish Laboratory*. Cambridge: Cambridge University Press, 2016.

Marks, S., P. Weindling, and L. Wintour (eds.). *In Defense of Learning: The Plight, Persecution and Placement of Academic Refugees, 1933–1980s*. London: British Academy, 2011.

Mendelsohn, E. *Heat and Life: The Development of the Theory of Animal Heat*. Cambridge, MA: Harvard University Press, 1964.

Meyerhof, O.F. "Phosphorus Metabolism." *American Scientist* 39, no. 4 (1951): 682–687.

Meyerhof, W. *In the Shadow of Love*. Santa Barbara, CA: Fithian Press, 2002.

Mulley, C. *The Woman Who Saved the Children: A Biography of Eglantyne Jebb, Founder of Save the Children*. Oxford: One World, 2009.

Nachmansohn, D. "Biochemistry as Part of My Life." *Annual Review of Biochemistry* 41 (1972): 1–30.

Nachmansohn, D. (ed.). *Metabolism and Function*. New York: Elsevier, 1950.

Nachmansohn D., S. Ochoa, and F.A. Lipmann. "Otto Meyerhof (1884–1951) A Biographical Memoir." *Science* 115, no. 2988 (April 4, 1952): 365–368.

Needham, D.M. *Machina Carnis: The Biochemistry of Muscular Contraction in Its Historical Development*. Cambridge: Cambridge University Press, 1971.

Norby, E. *Nobel Prizes and Life Sciences*. London: World Scientific, 2010.

Nye, M.J. *Blackett: Physics, War, and Politics in the Twentieth Century*. Cambridge, MA: Harvard University Press, 2004.

Prebble, J.N. *Searching for a Mechanism: A History of Cell Bioenergetics*. Oxford: Oxford University Press, 2019.

Przyrembel, A. "Friedrich Glum and Ernst Telschow." 2004. www.mpiwg-berlin.mpg.de/KWG/Ergebnisse/Ergebnisse20.pdf.

Rall, J.A. "Generation of Life in a Test-Tube: Albert Szent-Gyorgyi, Bruno Staub and the Discovery of Actin." *Advances in Physiology Education* 42 (2018): 277–288.

Rall, J.A. "Nobel Laureate A.V. Hill and the Refugee Scholars, 1933–45." *Advances in Physiology Education* 41 (2017): 248–259.

Rall, J.A. "The XIIIth International Physiology Congress in Boston in 1929: American Physiology Comes of Age." *Advances in Physiological Education* 40 (2016): 5–16.

Remy, S.P. *The Heidelberg Myth: The Nazification and DeNazification of a German University*. Cambridge, MA: Harvard University Press, 2002.

Robinson, A. *Einstein on the Run*. London: Yale University Press, 2020.

Sakmann, B. "Bernard Katz (1911–2003)." *Biographical Memoirs of Fellows of the Royal Society* 53 (2007): 185–202.

Skidelsky, R. *John Maynard Keynes: Hopes Betrayed.* Vol. 1. London: Penguin Books, 1994.

Smith-Rosenberg, C. *Disorderly Conduct.* New York: Oxford University Press, 1986.

Stadler, M. "Assembling Life: Models, the Cell, and the Reformations of Biological Science, 1920–1960." PhD diss., Imperial College, 2009.

States, D.M. "Otto Meyerhof and the Physiology Institute: The Birth of Modern Biochemistry." www.nobelprize.org/prizes/uncategorized/otto-meyerhof-and-the-physiology-institute-the-birth-of-modern-biochemistry/.

Sullivan, R. *Villa Air-Bel: World War II, Escape and a House in Marseille.* New York: HarperCollins, 2006.

Teich, M. *A Documentary History of Biochemistry 1770–1940.* Rutherford, NJ: Associated Universities Presses, 1992.

Tooze, A. *The Deluge.* London: Penguin, 2015.van der Kloot, W. *Great Scientists Wage the Great War.* Stroud: Fonthill Media, 2014.

Warwick, Andrew. *Masters of Theory: Cambridge and the Rise of Mathematical Physics.* Chicago: University of Chicago Press, 2003.

Werner, P. "Learning from an Adversary? Warburg against Wieland." *Historical Studies in the Physical and Biological Sciences* 28, no. 1 (1997): 173–196.

Werner, P. *Otto Warburg's Beitrag zur Atmungstheorie: Das Problem der Sauerstoffaktivierung.* Marburg: Basilisken-Presse, 1996.

Weston Smith, M. *Beating the Odds: The Life and Times of E.A. Milne.* London: Imperial College Press, 2013.

Wilson, D. *Rutherford: Simple Genius.* London: Hodder and Stoughton, 1983.

Wilson, T. *Churchill and the Prof.* London: Cassell, 1995.

Zimmerman, D. *Top Secret Exchange: The Tizard Mission and the Scientific War.* Montreal: McGill-Queens University Press, 1996.

Index